The Reuse and Recycling of CONTAMINATED SOIL

Stephen M. Testa
President
Testa Environmental Corporation
Mokelumne Hill, California

CRC Press is an imprint of the
Taylor & Francis Group, an **informa** business

CRC Press
Taylor & Francis Group
6000 Broken Sound Parkway NW, Suite 300
Boca Raton, FL 33487-2742

First issued in paperback 2020

ISBN-13: 978-0-367-57945-6 (pbk)
ISBN-13: 978-1-56670-188-4 (hbk)

Visit the Taylor & Francis Web site at
http://www.taylorandfrancis.com

and the CRC Press Web site at
http://www.crcpress.com

DEDICATION

To my parents
Joseph "Zip" Testa and Angelina Petitto Testa

AUTHOR

Stephen M. Testa is president and founder of Testa Environmental Corporation, located in San Juan Capistrano, California. He received his B.S. and M.S. in geology from California State University at Northridge. For the past 18 years, he has worked in the areas of geology, hydrogeology, engineering–environmental geology, and hazardous waste management with such firms as Bechtel, Converse Consultants, Ecology & Environment, Dames and Moore, Engineering Enterprises, Inc., and Applied Environmental Services, Inc.; the latter company he founded and served as its Chief Executive Officer.

Mr. Testa has participated in numerous subsurface hydrogeologic site characterization projects associated with nuclear and hydroelectric power plants, petrochemical and manufacturing plants, hazardous waste disposal facilities, and other industrial–commercial complexes. For the past five years, his main emphasis has been in the area of resource recovery, aquifer restoration, and soil reuse and recovery. Maintaining overall management and technical responsibilities in hygrogeologic, engineering geology, and hazardous waste-related projects, he has participated in numerous projects involving geologic and hydrogeologic site assessments, soil and water quality assessments, soil remediation, design and development of groundwater monitoring and aquifer remediation programs, waste minimization, soil reuse–recycling, resource recovery programs, and expert testimony.

He is the author of more than 60 technical papers and four books, including *Geological Aspects of Hazardous Waste Management, Restoration of Petroleum Contaminated Aquifers* (co-authored with Duane Winegardner), *Principles of Technical Consulting and Project Management* (co-authored with Duane Winegardner and Patrick Franks); he has also served as editor of and contributor to *Environmental Concerns in the Petroleum Industry.* Mr. Testa is an active member of numerous organizations, including the American Association for the Advancement of Science, Association of Engineering Geologists, American Association of Petroleum Geologists, Association of Groundwater Scientists and Engineers, Hazardous Materials Control Research

Institute, Geological Society of America, California Groundwater Association, and Sigma Xi. He is also a member of the American Institute of Professional Geologists (where he currently serves as President-Elect) and has served on the Executive Committee and various committees including those on professional development, continuing education, and screening. Mr. Testa conducts workshops on various environmental topics and has taught as an instructor at the University of Southern California and California State University at Fullerton.

PREFACE AND ACKNOWLEDGMENT

Over the past decade, the focus of the environmental industry and how we perceive what is considered a waste has progressively evolved. In the late 1970s and early 1980s, much attention was paid to the conduct of subsurface assessments in order to evaluate the magnitude of the environmental concerns associated with landfills, uncontrolled and abandoned industrial sites, and underground storage tanks, among numerous other structures and activities. These studies reflected upon how we as an industrial and manufacturing society managed materials now considered waste. By the mid-1980s, much attention was given to the recovery of light and dense nonaqueous phase liquid hydrocarbons (LNAPL and DNAPL, respectively) and the subsequent groundwater remediation and restoration of groundwater. During this same period, we also observed significant regulatory and technological advances in the area of soil remediation and treatment. Unfortunately, the technology or combination of technologies and regulatory strategies developed simply reduced the hazardous or toxic nature of the affected soil. The soil still had to undergo further handling, and in most instances, eventually had to be disposed at a landfill. When the removal and excavation of contaminated soil was neither technically feasible nor cost effective, *in situ* approaches were developed. With the high costs and liabilities associated with materials considered toxic or hazardous waste, industry has continually worked toward a means to minimize the volume of waste being generated and to reclaim, reuse, or recycle materials that are or were once considered waste.

As a practicing professional engineering and environmental geologist for the past 20 years, I have observed these transitions first-hand. My first encounter with the concept of resource recovery was while working on numerous refinery sites, being primarily involved with the recovery of huge volumes of LNAPL that had originated from a variety of sources and been released into the subsurface over the past 50 years or more. Some of the sites investigated had enough recoverable LNAPL to offset the cost of recovery. It was clear in the mid-1980s, however, that such sites, as well as others, had accumulated huge volumes of contaminated soil that would eventually need to be addressed. Many refineries, tank farms, and other industrial–manufacturing facilities were typically characterized by large volumes of contaminated soil on site and by large areas of unpaved surfaces in need of pavement or areas of existing pavement and land surfaces requiring reconditioning; as a result, we began to rethink how we view contaminated soil. It is my view that much of what we refer to as contaminated soil should be considered a resource that can be put to beneficial use.

Many materials once considered waste are presently being reused and recycled as ingredients in some industrial or manufacturing processes to produce commercially viable products. Materials such as rubber tires, fly ash, slag, glass and plastics, have been utilized in such a manner almost exclusively by the asphalt and concrete industries. Just during the past few years, there has been much interest in the reuse and recycling of contaminated soil. Despite this interest and current practice, such programs are regionally localized and are often overlooked, with the parties involved unaware that such alternatives to soil remediation, treatment, and/or disposal exist. As increasing awareness of reuse and recycling technologies as applied to contaminated soil becomes more prevalent, it is important that both environmental and engineering concerns be adequately addressed. Specifically, it is necessary to adequately evaluate how contaminated soil should be handled and processed, what waste materials can be incorporated, what quantity of waste can be incorporated while still adhering to the specifications set forth for the end use, how the resulting end product should be evaluated for performance and leachability, and how to incorporate contaminated soil in a cost effective manner in order to provide financial incentives and benefits for implementation of such programs. In addition, it is important that such programs minimize the liability associated with being a generator of hazardous waste to its least denomination. This book attempts to address these objectives.

This book focuses on the soil problem and presents a new way of viewing contaminated soil — as not a waste but rather a resource that in many instances can be recovered. The book is divided into 12 chapters. Chapter 1 introduces the reader to the soil problem and the concepts of reclamation, use, reuse, recycling, and resource recovery. Discussions of some of the various waste types suitable for reuse and recycling are also presented. Chapter 2 discusses the more pertinent regulations and regulatory programs relevant to the reuse and recycling of contaminated soil. Reuse and recycling technologies are presented in Chapter 3, including descriptions of the processes involved and discussion of their respective advantages, limitations, and relative costs. Practices and considerations related to performance of certain field activities are presented in Chapter 4. Laboratory considerations, including discussions of waste types and characterization, and acceptance criteria are presented. The concept of leachability and discussions of the various leachability tests currently available are also provided. Chapter 6 presents engineering considerations and applicable tests to be considered on the preprocessed soil and produced product. In an effort to understand the physical and chemical mechanisms involved, Chapter 7 presents discussions of the chemical aspects of bitumens and asphalt, including bitumen types and asphalt chemistry. Factors important to the incorporation of contaminated soil into asphalt, such as stability, contaminant mobility, durability, aging, biological resistance, permeability, and leachability are also discussed. Asphalt emulsions or binders in regard to types, chemistry, production and specifications, breaking and curing, tests, and selection are presented in Chapter 8. Chapter 9 discusses the chemical aspects of cementitious materials. Included is discussion of the general nature of cement chemistry and those physical and chemical factors important to the incorporation of contaminated soil, such as strength, durability, and contaminant immobilization and leachability; degradation and biological resistance are also discussed. The types of contaminants suitable for incorporation as an ingredient to produce commercially viable products and some examples of their application and utilization are presented in Chapters 10 and 11, respectively. Case histories on the incorporation of contaminated soil into asphaltic end products appear in Chapter 12.

This book could not have been completed without the devoted assistance of my wife, Lydia, who is responsible for typing the majority of this book, and Ms. Coleen Burke-Franco, my executive assistant. I was assisted in the time-consuming effort of performing literature searches and retrieval of pertinent published papers, reports, and documents by Ms. Beverly Dowdy and Ms. Coleen Burke-Franco. I have also been aided by fruitful discussions with Dr. Jim Conca and Duane Winegardner and persistent encouragement from Robert L. Thatcher and Jerry Blodgett. Any errors contained herein are solely credited to the author.

While I was in high school, my parents were responsible for landing my first full-time job, as a raker on a hot-mix asphalt crew in the San Fernando Valley one very hot and smoggy summer, for which I earned $1.17 an hour. I hated this job and promised myself I would never get involved with asphalt again. For whatever appreciation of asphalt gleaned from this early and brief encounter with the stuff, I can only thank my parents, to whom I dedicate this book.

Stephen M. Testa

CONTENTS

LIST OF TABLES

LIST OF FIGURES

1 INTRODUCTION

1.1 THE SOIL PROBLEM

The United States, along with almost every other country, is experiencing a continuing increase in the amount and type of materials being discarded, which places a significant burden on landfills and disposal sites, and increased costs for business owners and industry everywhere. In 1960, the annual amount of solid waste produced in the United States was about 82 million metric tons (90 million tons). In 1986, this annual amount increased to 146 million metric tons (161 million tons) and to 164 million metric tons (181 million tons) in 1988. In 1993, the amount was 4.1 billion metric tons (4.5 billion tons; Figure 1-1). Over the past two decades, contaminated soil has made up a continually increasing amount and significant volume of this material. This situation is exacerbated by the physical, economic, and technical limitations associated with the technologies currently available for the remediation of contaminated soil.

Soil contamination is a major concern not only throughout the United States but worldwide. The impact on soil from industrial and agricultural practices, management of Superfund sites, exploration and production, and mining and nuclear industrial practices, among others, remains difficult to assess. Certainly petroleum-contaminated soil affects the largest number of sites and the largest total volume of impacted material. The volume of petroleum-contaminated soil that is either discovered or is generated each year is not consistently tracked on a state-by-state basis, so this total is unknown. The overall amount of contaminated soil generated, however, can be staggering. For example, in some states such as Oklahoma, contaminated soil accounts for about 98% of the waste generated as a one-time occurrence. In many cases, the volume of contaminated soil generated at a particular site or the remedial technology used is not consistently monitored and may not be monitored at all. Some perspective, however, can be gleaned from various industry evaluations conducted over the past few years and from information generated as part of the underground storage tank (UST), Superfund, and other regulatory programs (i.e., reuse/recycling programs).

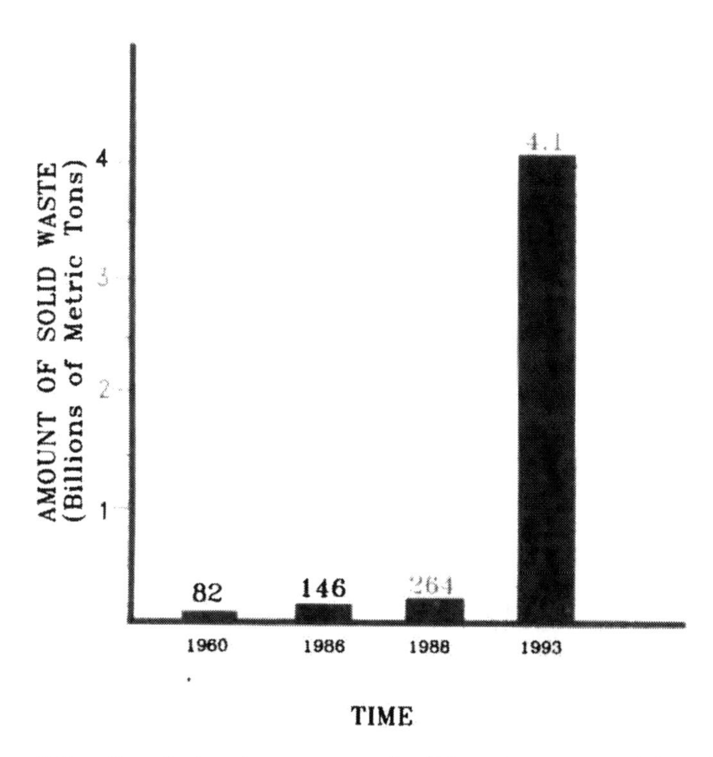

Figure 1-1. Graph showing amount of solid waste generated since 1960.

1.1.1 Underground Storage Tanks (USTs)

Petroleum contamination is commonly associated with USTs (Figure 1-2). Millions of USTs were installed in the 1950s and 1960s. The U.S. EPA estimates that out of an estimated 2.5 to 3 million USTs throughout the United States, there are more than 400,000 USTs that have leaked or are leaking petroleum hydrocarbons, with more expected to develop leaks in the future. If one assumes that an average estimated amount of contaminated soil as a result of such leaks is on the order of 50 to 80 yd^3, then the volume of contaminated soil solely attributable to USTs is on the order of 20 to 32 million yd^3. This number, however, is very conservative since in many cases much larger volumes ranging up to thousands of cubic yards per site have been reported on numerous occasions, reflecting significant releases. In addition, thousands of USTs remain unrecorded and their individual impacts on the subsurface remain unknown.

Although all states have UST programs in place, most states can only provide an estimate of how petroleum-contaminated soils from UST sites are handled, with only a select few having actual statistics. A random look at a few states clearly demonstrates this.

Florida maintains no database which tracks contaminated soil, although most of such soil is treated *in situ* via bioremediation or soil vapor extraction.

Figure 1-2. Removal of an old UST with suspected leakage.

What is not treated *in situ* is excavated and subsequently undergoes thermal treatment. Less than 25% of what is excavated goes into landfills, and none of such material is recycled.

In Georgia, processors must recycle 60% in total volume within the first 90 d, and of the 40% that remains, 60% of it must be further recycled. Most of what is thermally treated is utilized as fill material, sub-base, or is incorporated into asphalt.

In Illinois, as in many other states, the vast majority of contaminated soil goes into municipal landfills. Records are kept for the volume of contaminated soil that is generated at each site, but no totals are maintained. The amount generated per site is on the order of 75 to 1470 yd^3, averaging about 608 yd^3. In 1994, 2.3 million yd^3 of contaminated soil was excavated from UST sites, of which 100,000 yd^3 treated by bioremediation, vapor extraction, or thermal treatment, and only a minor amount of about 5% was recycled.

Indiana is just beginning to track petroleum contaminated soil. However, no such records are maintained by such states as New York, Michigan, and Washington.

About 45% of petroleum-contaminated soil is thermally treated in Minnesota, with about 40% of land treated; the remainder, about 15%, undergoes assorted options, such as biofiles. The majority of the thermally treated material is eventually incorporated into asphalt. In 1994, about 17,706 yd^3 of petroleum-contaminated soil was excavated from 171 UST sites. Of this total, 2875 yd^3 from 22 sites underwent land treatment, whereas 6420 yd^3 from 41 sites underwent thermal treatment. About 4842 yd^3 from four sites underwent composting, while about 720 yd^3 from 14 sites was either disposed of in landfills, treated outside Minnesota, used as road embankment, soil vented, or thinly spread on the ground surface. About 210 tons of asphalt were processed and 3683 tons of roadbase were produced following thermal treatment.

Ohio recently commenced tracking the volume of contaminated soil generated and its ultimate treatment in 1995. The vast majority of soil, 90% or greater, goes to landfills or is bioremediated in petrocells. There are currently two or three petrocells in Ohio, with some recent request for landfarming.

In Pennsylvania, an estimated 3000 to 4000 UST sites are recorded. The vast majority of contaminated soil goes into landfills, although the trend is toward on-site treatment, with virtually no recycling performed at the time of this writing.

In Wisconsin, about 50% of petroleum-contaminated soil is treated *in situ*. The remainder is excavated, of which 50% ends up in landfills and the rest is remediated via thermal treatment or bioremediation. In 1992, asphalt incorporation accounted for approximately 7% of contaminated soil excavated, but in 1993 accounted for less than 3%, with an increase in thermal desorption from 18 to 22% during the same period. Asphalt incorporation, bioremediation, thin spreading, and land spreading were the least used technologies in 1992 and 1993, accounting for less than 17% of the volume of contaminated soil excavated. For *ex situ* soil, landfill disposal is the most common option used. With respect to both the number of sites and the volume of material disposed, landfills accounted for about 61 to 67% of soil excavated at UST sites, and 26 to 32% in volume was remediated via landfill disposal.

The current awareness of petroleum hydrocarbon-affected sites from USTs, however, barely scratches the surface when one considers the volume of soil impacted by petroleum and other hydrocarbon constituents and metals at industrial and manufacturing complexes such as refineries, tank farms, foundries, power plants, and auto wrecking yards. The volume of petroleum hydrocarbon-contaminated soil at a moderate-sized refinery, for example, can range up to millions of cubic yards.

1.1.2 Chemical Sites

A 1979 survey of 53 of the largest chemical manufacturing companies in the United States, conducted for the period 1950 to 1979 and referred to as the Eckhardt survey, reported almost 17 million tons of organic generated waste disposed (Table 1-1). Of this total, a little over 10 million tons were untreated (i.e., in landfills, ponds, lagoons, and injection wells). A little less than 0.5 million tons were incinerated, and a little over 0.5 million tons were either recycled or reused. Not addressed is the volume of contaminated soil generated as a one-time occurrence, as is typical of any remediation activity.

As of 1986, about 68,265 generators of Resource Conservation and Recovery Act (RCRA)-regulated waste were identified by the U.S. EPA Hazardous Waste Data Management System. The largest portion of federally regulated waste generators was in EPA Region 5, which includes the industrial states of Illinois, Indiana, Michigan, Minnesota, Ohio, and Wisconsin. Manufacturing

Table 1-1 Amount of Waste Disposed per Disposal Method

	Untreated				Treated						
SIC[a] Division	Landfills	Pits, ponds, lagoons	Injection well	(Subtotal)	Incinerated	Recycled or reused	Evaporated	(Subtotal)	Unknown	Others[b]	Total
Organics	2,200	5,237	3,225	10,662	485	539	48	1,072	42	5,067	16,843
Inorganics	2,568	4,702	2,584	9,854	357	77	5,856	6,290	239	3,798	20,181
Plastics	1,112	254	322	1,688	115	385	7	507	11	559	2,765
Agricultural	1,145	351	708	2,204	912	39	2,328	3,279	179	10,686	16,348
Misc. chemicals	132	122	10	264	19	27	10	56	1	8	329
Other	986	4,240	170	5,396	60	60	332	432	0.65	12	5,861
Medicinal and soaps	122	9	0.96	132	85	0.65	0	86	0	135	353
Unknown	33	4	5	42	3	0	179	182	0.90	120	345
Total	8,298	14,919	7,025	30,242	2,036	1,128	8,760	11,924	474	20,385	63,025

Note: Data is presented in units of thousands of tons.

[a] Standard Industrial Classification.

[b] Includes storage, land applications, and ocean disposal, among other methods.

Modified from PEDCO (1979).

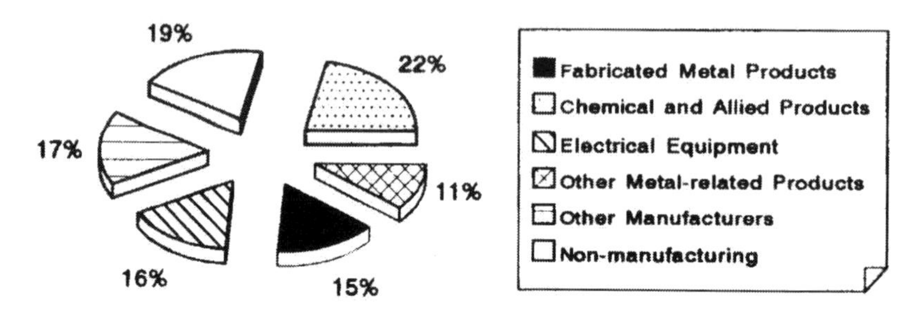

Figure 1-3. Number of hazardous waste generators per industry type.

industries account for approximately 86% of the generators (Figure 1-3). In 1981, the chemical and petroleum industries accounted for an estimated 71% of the hazardous waste generated and possibly up to 85% of the total quantity of the waste generated. Metal-related industries generated 22% of the waste, with other industries accounting for the remaining 7%.

1.1.3 Oil Field Sites

Another source of rather large volumes of hydrocarbon-contaminated soil is related to production in past and existing oil fields. From a regional industry perspective, the impact of oil exploration and production activities has not been highly visible, especially in areas outside of urbanized areas far from public scrutiny or in areas of the Midwest such as Oklahoma and Texas where it is still relatively inexpensive to dispose of crude and petroleum hydrocarbon-contaminated soil in comparison to *in situ* or above-ground treatment, or reuse and recycling.

This is not, however, the case everywhere, as is evident within the densely populated and urbanized southern California, area. In southern California oil fields that reach the end of their productive lives are rapidly taken out of production and the land redeveloped. The rich history of oil and gas exploration and exploitation in California goes back to the 1860s, the first years of commercial production. California comprises 56 counties, of which 30 are known to produce petroleum. Oil is chiefly produced in 18 of these counties, notably within the Los Angeles, Ventura, San Joaquin, and Santa Maria basins. Within the vicinity of the heavily congested city of Long Beach, located in Los Angeles County, approximately 12 mi^2 or 3840 acres are currently, or historically have been, petroleum-producing properties. The remaining 12 counties produce gas. In these areas, disposal of contaminated soils has emerged as a serious problem.

Potential primary sources of soil contamination include oil wells, sumps, pits and dumps, leakage from above-ground storage tanks and containers, reservoirs, and random spillage or leakage. Secondary sources include pumping stations, piping ratholes, transformers, underground storage tanks, and well cellars. The primary and most common compounds and constituents associated

Table 1-2 Crude Oil Criteria in Consideration as a Hazardous Waste

Parameter	Maximum contaminant level (parts per million)
Arsenic	5
Chloride	3,000
Chromium	10
Flash point	< minimum standard[a]
Lead	50
Polychlorinated biphenols (PCBs)	5

[a] Less than minimum standard as set by ASTM for recycled products.

with oil field properties that may be considered hazardous include methane, crude oil and associated constituents, and drilling mud and associated constituents. A summary of general characteristics of crude oil for consideration as hazardous waste is presented in Table 1-2.

In addition to the presence of oil fields, numerous high-volume petroleum-handling facilities and associated refineries, terminals, and pipeline corridors exist in close proximity to production fields and residential areas (Figure 1-3). Constituents and compounds of concern at these sites may include crude oil or refined products, including volatile organic compounds. All of these sites, both historically and in the present, contribute to the overall volume of contaminated soil generated.

1.1.4 Geothermal Sites

Geothermal operations, as with oil field properties, provide another example of the close relationship between site usage and the potential for adverse environmental impact. Geothermal plants typically comprise a power plant, brine storage ponds, drill sumps, and possibly a leach field. Constituents associated with geothermal operations that may be considered hazardous, fall into two categories: brine and lead-mine scale.

Brines are geothermal mineralizing fluids composed of warm to hot saline waters containing calcium, sodium, potassium, chloride, and minor amounts of other elements. Line-mine scale is the calcified material that forms within the interior of the process piping system associated with geothermal steam generating operations. Geothermal wells are drilled into a superheated, water-bearing strata. Superheated water and steam are recovered from these wells and directed through a process piping and condensed as a power source to drive the turbines of the electrical generators.

Various mineral and elemental metal compounds exist naturally in the superheated water and steam; these cause calcification to occur in the form of a scale which builds up on the interior of the process piping. Unrestricted buildup of this scale inhibits the operation of the units by inhibiting the flow of water and steam to the turbines, necessitating costly shutdown and line cleaning programs. To overcome this situation, a process called line-mining is used. Wire mesh rolls are inserted within the process piping, in conjunction with alteration of the water's pH levels. The scale buildup is then directed to the mesh rolls and away from the piping walls. The process water flow is thus literally mined of the minerals and metal compounds as the lead scale builds up on the wire mesh.

The chemicals of concern associated with the operation, abandonment, or redevelopment of such sites include arsenic, copper, lead, and zinc. General representative chemical characteristics for both brine and lead-mine scale are presented in Table 1-3.

1.1.5 Manufactured Gas Plants

Manufactured gas plants (MGPs) have operated since the late 1890s, with many industrial sites having undergone redevelopment. Recent surveys estimate that about 3000 MGP sites exist throughout the United States. Associated with such sites are massive volumes of tarry materials, which are produced by several coal gasification processes including coal carbonization (CC), caburetted water gas (CWG), oil and natural gas (OG), or a combination thereof (i.e., CC and CWG, or OG and CWG). By the 1960s, natural gas or, to a lesser degree, oil gas processes have been used.

Tars are commonly known to contain a wide variety of organic and inorganic compounds, including monocyclic and polycyclic aromatic hydrocarbons (MAHs and PAHs, respectively), compounds that are known or suspected human carcinogens. Tars can be produced from different processes and stored in a variety of ways. They can be quite varied with respect to their physical and chemical characteristics, although some chemical similarities are prevalent, particularly with respect to the type and abundance of certain constituents, notably PAH compounds. It is estimated that the average volume of tar-contaminated soil and associated groundwater and surface water averages 10,000 yd^3 per site. Free tar and oil at individual sites are estimated to range from about 1 to 50,000 gal from residues left in holding tanks, ponds, and trenches. A summary of residuals from gas manufacturing and processing of coal, coke, and oil and the chemical characteristics of such materials are presented in Table 1-5.

1.1.6 Mining Sites

Mining on a grand scale commenced in the mid-1800s in such mining districts as Leadville, Colorado, the Tri-State mining area of Oklahoma, Kansas and Missouri, and Anaconda, Montana, among others. Since the 1940s, the exploration and production of uranium and other strategic minerals have

Table 1-3 Summary of General Characteristics of Brine and Lead-Mine Scale

	Concentration		
		Scale (in wt %)[a]	
Parameter	Brine (mg/l)	Well head separator scale	Injection line scale
Aluminum	ND (<2)	0.3–0.72	ND
Antimony	ND (<2)	NA	NA
Arsenic	1.39–2.7	NA	NA
Barium	680–1,520	NA	NA
Cadmium	1.3–340	NA	NA
Calcium	12,700–34,000	0.2–0.77	0.58–0.81
Cesium	17–22	NA	NA
Chloride	80,200–184,000	NA	NA
Chromium	ND (<0.1)	NA	NA
Cobalt	0.2–0.3	NA	NA
Copper	1.7–5.9	0.1–1.6	ND
Iron	490–4,540	7.7–53.8	29.0–37.3
Lead	17–300	15.7–40.6	NA
Lithium	210–320	NA	NA
Magnesium	100–200	0.1–0.75	0.1–0.17
Manganese	470–1,860	0.2–0.5	0.47–0.68
Mercury	ND (<0.004)–0.001	NA	NA
Molybdenum	ND (<0.2)	NA	NA
Nickel	ND (<0.2)	NA	NA
Potassium	6,900–15,400	NA	NA
Rubidium	34–83	NA	NA
Selenium	270–320	NA	NA
Silicon	ND (0.6)–0.1	0.55–8.0	19.5–41.70
Silver	<1–1	2.4–6.0	2.12
Sodium	34,800–68,000	NA	NA
Strontium	640–1,510	NA	NA
Sulfur	NA	5.70–18.1	ND (0.05)
Total dissolved solids	128,000–262,000	NA	NA
Tin	ND (<10)	NA	NA
Zinc	190–1,170	0.6–7.2	NA
Anions			
Ammonia	69–150		
Bromide	107,000–190,000		
Chloride	ND (<1)		
Fluoride	ND24–ND (<30)		
Iodide	ND (<5)		
Nitrate	ND (<6)		
Sulfate	53–60		
TDS	559–874		
Thiosulfate	1.1–1.199		

Note: ND, Not detected at analytical detection limit shown in parentheses; NA, not available.

[a] Concentration in mg/kg.

Table 1-4 Summary of Anticipated Residuals from Gas Manufacturing Processing of Coal, Coke, and Oil

Process residuals	Physical form and principal chemical content	Gas manufacturing process		
		Coal carbonization	Carburetted water gas	Oil gas
Ammonia liquors	Aqueous liquid: inorganics, phenolics	X	—	—
Ash and clinker	Solid: metals (and unburned coke or coal)	X	X	—
Carburetted water gas tar	Liquid: PAHs, BETX	—	X	—
Coal tar	Liquid: PAHs, BTEX, phenolics	X	—	—
Coke and coke breeze	Solid: pyrolyzed coal	X	—	—
Lampblack	Sludge: elemental carbon and oil tar	—	—	X
Oil tar	Liquid: PAHs, BTEX	—	—	X
Spent oxide or lime, wood chips (support media)	Solid: metals, cyanide, sulfur, tar	X	X	—
Tar sludges	Solid/liquid: PAHs, BTEX	X	X	X
Tar/oil/water emulsions	Aqueous and organic liquids: PAHs, BTEX	X	X	X
Wastewater treatment sludges	Solids, aqueous and organic liquids: inorganics, phenolics, PAHs, BTEX	X	X	X

Note: X indicates residual was produced; — indicates residual was not produced in substantial amounts; PAH = polycyclic aromatic hydrocarbons; BTEX = benzene, toluene, ethylbenzene, and xylene.

Table 1-5 Chemical Characteristics of Free Tars and Oil Associated with MGP

Parameter	Sample Number							
	1	2	3	4	5	6	7	8
Metals								
Arsenic	20	6.4	7.8	9.3	23	4.7	3.0	3.6
Beryllium	<1.0	<1.0	<1.0	<1.0	<1.0	<1.0	<1.0	<1.0
Cadmium	<1.0	<1.0	1.2	<1.0	<1.0	3.7		4.4
Chromium	1.1	11	28	1.8	36	36		2.4
Lead	1.0	50	44	1.4	930	8.1		37
Nickel	2.1	74	52	2.2	5.1	5.6		1.9
Selenium	1.7	1.1	3.2	1.9	4.5	<1.0		4.0
Vanadium	6.9	230	27	5.8	12	14		7.6
Others								
Cyanide	<1.0	<1.0	4.5	<1.0	<1.0	26	1.7	130
MAHs								
Benzene	550	460	14.00	4.00	4,970	11	790	450
Toluene	2,120	1,050	9.00	53.0	7,000	<1.00	1,310	1,060
Ethylbenzene	1,860	450	37.0	300	2,700	<1.00	400	170
Xylene (total)	5,120	1,450	158	2,390	8,060	2.00	1,050	1740
Styrene	60	100	63	1,060	2,490	<1.00	140	200
PAHs								
Indan	3,800	1,390	640	410	ND	ND	ND	ND
Naphthalene	70,700	13,300	4,030	52,000	135,000	970	6,540	21,400

Table 1-5 Chemical Characteristics of Free Tars and Oil Associated with MGP *(continued)*

Parameter	Sample Number							
	1	2	3	4	5	6	7	8
2-Methylnaphthalene	33,400	7,450	3,080	37,400	34,100	700	2,080	5,770
1-Methylnaphthalene	23,500	4,900	2,350	21,500	19,600	640	1,170	3,320
Acenaphthylene	610	260	300	12,100	17,800	890	1,180	530
Acenaphthene	11,900	340	1,150	900	1,650	460	270	500
Dibenzofuran	4,000	170	120	1,090	1,150	1,850	820	1,650
Fluorene	11,600	1,350	1,110	6,870	8,950	1,420	1,080	2,100
Phenanthrene	32,600	5,210	3,470	17,000	34,600	11,300	3,130	7,000
Anthracene	8,570	390	690	5,420	7,670	2,980	980	1,080
Fluoranthene	13,400	1,500	1,360	4,760	12,600	8,520	1,900	4,150
Pyrene	13,200	2,410	2,070	6,890	17,900	6,170	1,000	3,080
Benz(a)anthracene	4,900	750	450	2,340	5,130	1,920	610	1,470
Chrysene	5,100	1,050	750	2,340	6,470	2,360	620	1,880
Benzo(b)fluoranthene	2,150	1,240	270	1,350	3,130	1,540	490	1,150
Benzo(k)fluoranthene	2,950	1,050	130	610	2,850	830	360	850
Benzo(a)pyrene	3,900	1,820	390	1,560	6,420	1,410	500	1,330
Indeno(1,2,3-cd)pyrene	2,610	1,400	690	1,080	4,890	3,500	790	1,720
Dibenz(a,h)anthracene	490	ND	250	140	640	450	180	350
Benzo(g,h,i)perylene	3110	1640	940	1270	7090	4150	960	2290

Note: All data in mg/kg; ND = not detectable at or exceeding the parameter's respective analytical detection limit.

Modified after Electric Power Research Institute, 1993, Chemical and Physical Characteristics of Tar Samples from Selected Manufactured Gas Plant (MGP) Sites, EPRI TR-102184, May, 1993.

also left a legacy of uranium mill tailing piles, asbestos- and metal-contaminated soil, and degradated surface water and groundwater.

Mining sites can vary in size from less than 1 acre to several square miles. Large amounts of waste can be produced, and virtually all such waste generated is disposed of on-site. Overburden and waste rock are typically placed in piles adjacent to the mine or used as backfill. Tailings are commonly disposed of as a slurry in unlined tailing impoundments, covering some 1500 ha within 24 sites throughout the western states. Primarily derived from extensive mining activities of the 1950s and 1960s, the material is composed of finely comminuted uranium ore in which all the uranium has been removed, but the radioactive daughter elements produced by radioactive decay remain. The most important of the radioactive daughter elements is radium (^{226}Ra). If ingested, Ra can cause damage due to its intense alpha radiation. Ra also decays to produce radon (^{222}Ra), a radioactive gas which can further decay into suspended solid radioactive products such as lead (^{210}Pb) which, if inhaled, can lead to cancer.

The U.S. EPA has estimated that about 31 trillion kg of mine waste and 13 trillion kg of tailings accumulated between the years 1910 and 1981, with over 80% of the overburden and tailing wastes generated by the phosphate rock, uranium, copper, and iron ore mining segments of the industry. Uranium mill tailings currently total more than 230 million tons at mill sites throughout the United States. Out of 1174 sites that are listed on the National Priority List (NPL) under Comprehensive Environmental Response, Compensation, and Liability Act (CERCLA), a minimum of 109 sites contain mining wastes as a result of past or current mining activity or as a result of the disposal of mining waste on-site.

Mining waste is defined as those wastes generated from the mining, milling, smelting, and refining of ores and minerals. Mining specifically refers to the extraction or removal of ores, whereas milling refers to the concentration of extracted ores via crushing, screening, or chemical treatment of the ore rock. Primary smelting or refining reflects further processing of the ore product. Mining wastes include soil and rock (overburden) that is removed to expose an ore body in a surface mine. Rock that is excavated during surface open-pit mining or underground mining operations includes overburden, waste rock, tailings, slag piles, and heap leach residue derived from dump and heap leach operations and waste water.

Overburden is the surficial soil and rock that is stripped from the area overlying a particular ore body. These materials are typically chemically inert and are not usually classified as hazardous. Waste rock refers to the non- or low-mineralized rock that has been removed from above or adjacent to the ore body. Waste rock is usually stockpiled in close proximity to the mining operation. Tailings are typically uniform, finely ground rock particles derived at the beneficiation and extraction plant after the commercial ore has been extracted. The potential for waste rock or tailings to be of environmental concern will depend on the nature and extent of the mineralization, the overall

Table 1-6 Typical Mining Wastes and Associated Mining Activities

Waste type	Mining activity
Acid	Oxidation of naturally occurring sulfides in mining waste, notably copper, gold, and silver
Asbestos	Asbestos mining and milling operations
Cyanide	Precious metals heap-leaching operation
Leach liquors	Copper-dump leaching operations
Metals (Pb, Cd, Ar, Cu, Zn, and Hg)	Mining and milling operations
Radionuclides (radium)	Uranium and phosphate mining operations

chemistry of the materials, climatic conditions, and the buffering capacity of the underlying soil.

Heap leach residue is the crushed and agglomerated ore piles that have undergone leaching by the passing of solutions through the materials. Prior to closure, the ore is reused and possibly treated to meet specific residual concentration levels. Heap leach residue may be considered as hazardous waste, depending on the ore chemistry reagent types and the extent of leaching and/or treatment. Waste water is generated during performance of flotation, acid leaching, or solvent extraction processes or from operational activities such as boiler operations, residual spent leach solutions, and washdown. Overburden, waste rock, tailings, and heap leach residue are all potential candidates for reuse and recycling.

Typical mining waste types and associated mining activity are presented in Table 1-6. The majority of the contaminants are metals such as arsenic, cadmium, copper, lead, and zinc. Nonmetals include mercury, asbestos, and radium.

Although some form of mining is conducted in every state, most metal mines are situated in the western United States with similar commodities usually restricted to certain geologic environments. Twenty-five states are chief producers of copper, gold, iron, lead, phosphate, silver, and zinc. Major producing states, the quantity of waste produced, and the environmental concerns associated with mining activity are presented in Table 1-7.

All solid waste management practices pertaining to the mining industry were investigated by RCRA in 1976. The U.S. EPA evaluated the sources and volume of mining waste generated, disposal practices, potential danger, alternative disposal alternatives and their cost, and potential uses of the waste. When RCRA was amended in 1980, the Bevill Amendment directed the U.S. EPA to conduct a comprehensive study to prohibit regulation but to assess the adverse environmental effects, if any, of the disposal and utilization of solid wastes from the extraction, beneficiation, and processing of ores and minerals. Furthermore, the U.S. EPA was required to state whether regulation of mining wastes was warranted and which regulations were appropriate. In 1986, the

Table 1-7 Summary of Major Mining Areas of Environmental Concern

Ore	No. of sites	Quantity of wastes (million metric tons/yr)	Environmental concern	Principal producing states[a]
Copper	19	502	Elevated metals, low pH	Arizona, New Mexico, Utah
Gold and silver	117	100	Radioactivity	Nevada, Montana, California
Iron	26	177	Elevated metals, low pH	Minnesota, Michigan, Missouri
Lead and zinc	23	18	Residual cyanide	Missouri, Tennessee, Idaho
Phosphate	27	403	Elevated metals, low pH	Florida, North Carolina, Idaho
Other metals	24	62	Elevated metals, low pH	Colorado (molybdenum)
Total	226	1262		

[a] Source: Bureau of Mines Yearbook (1985).

EPA determined that RCRA standards would likely be unnecessary, technically infeasible, and economically impractical when applied to mining wastes. Mineral wastes were subsequently excluded from regulation under RCRA Subtitle C, since these wastes were considered raw materials used in the production process or product. Only leach solutions that escape from the production process and abandoned leach piles were considered wastes. In 1988, six processing wastes were classified as hazardous and excluded from the Bevill Amendment, and 15 high-volume mineral processing wastes that would soon come under the amendment were identified. All other processing wastes would be excluded from the amendment.

1.1.7 Environmental Terrorism

Prior to 1991, very little concern existed with the general public regarding the potential for environmental terrorism on a regional scale. Certainly the Gulf War proved otherwise. After the Gulf War, 700 of Kuwait's 1500 oil wells were releasing oil into the environment; 600 of these wells were aflame. An estimated 11 million barrels of oil were being burned or spilled each day; by August this estimate was reduced to 6 million barrels following the capping of 261 wells. In addition to harmful effects to the atmosphere and the Persian Gulf, numerous pools of oil formed. Some of these pools were reported to be up to 4 ft deep and collectively contained an estimated 20 million barrels of oil.

The presence of the oil pools is most likely the larger long-term problem. The overall result is huge volumes of crude oil and petroleum-contaminated soil that eventually will need to be addressed (Figure 1-4).

1.2 REMEDIATION VS. REUSE AND RECYCLING

Over the past decade, numerous technologies or combinations of technologies have been used to remediate contaminated soil. Some of the more conventional approaches to the remediation of petroleum hydrocarbon-contaminated soil, excluding *in situ* technologies such as vapor extraction and bioremediation, include dig-and-haul followed by direct disposal, thermal destruction, bioremediation, chemical fixation, and incineration. The technologies available for the remediation of soil contaminated by various metals such as lead are more restrictive and consist predominantly of fixation-type processes and direct disposal. A compilation of various soil remediation technologies is presented in Table 1-8.

Although many advances and innovations have occurred in these areas, most of these processes do not alleviate a producer's liability as a generator of hazardous waste, nor do they meet the criteria of reuse and recycling. Simply reducing the toxicity of the material, thus making it easier to handle or making it possible to reduce the category of landfill where the material will ultimately be disposed, is not reuse or recycling. Utilizing the waste, however, in such a

Figure 1-4. Al-Rawdhatayn oil field showing environmental impact of Gulf War on desert environment as of March, 1991. Surficially stained desert surface (top) and buildup of berms to contain oil (bottom) is shown.

Table 1-8 Summary of Soil Remediation Technologies

Technology	Applicable contaminants
	In situ
Volatilization (vapor extraction)	Volatile organic compounds
Bioremediation	Petroleum hydrocarbons[a]
	Chlorinated solvents
Leaching/chemical reaction	Petroleum hydrocarbons
	Chlorinated solvents
Vitrification	Petroleum hydrocarbons
	Chlorinated solvents
	Coal-tar residues
Stabilization	Petroleum hydrocarbons
	Chlorinated solvents
	Coal-tar residue
	Metals
	LLW and ILW
Passive	Petroleum hydrocarbons
	Chlorinated solvents
	Coal-tar residues
Isolation/containment	Petroleum hydrocarbons
	Chlorinated solvents
	Coal-tar residues
Natural attenuation	Petroleum hydrocarbons
	Chlorinated solvents
	Not *in situ*
Land treatment	Petroleum hydrocarbons[b]
	Coal-tar residues
Thermal treatment	Petroleum hydrocarbons
	Chlorinated solvents
	Coal-tar residues
Solidification/stabilization	Petroleum hydrocarbons
	Chlorinated solvents
	Coal-tar residues
	Metals
Chemical extraction	Petroleum hydrocarbons
	Chlorinated solvents
	Coal-tar residues
	Metals
Excavation/disposal	Petroleum hydrocarbons
	Coal-tar residues
	LLW and ILW
	Metals

[a] Excludes VOCs.

[b] Petroleum hydrocarbons include gasoline and fuel oils such as diesel and kerosene.

manner that the end result is a viable commercial product distinguishes reuse and recycling technologies from treatment and remediation technologies.

The emphasis in waste management over the past few years has changed from treatment (including *in situ* treatment and risk assessment) and disposal to waste minimization. Waste minimization is the feasible reduction of hazardous waste that is generated or subsequently would be treated, stored, or disposed. This includes any source reduction or recycling activities performed by a generator that results in the reduction of the total volume or quantity of the waste and/or the reduction of the toxicity of the waste. Pollution prevention at its source, which has been used synonymously with pollution prevention, differs in that it also addresses air emission, water discharges, and solid wastes as hazardous waste streams. Pollution prevention also attempts to avoid the potential transfer of risk from one medium to another. A schematic flow-chart illustrating the hierarchy of waste management alternatives is shown in Figure 1-5.

Certainly the initial objective or first option in waste management practices is source reduction, which is the elimination or reduction of waste at the source. Source reduction measures can occur via operational changes that

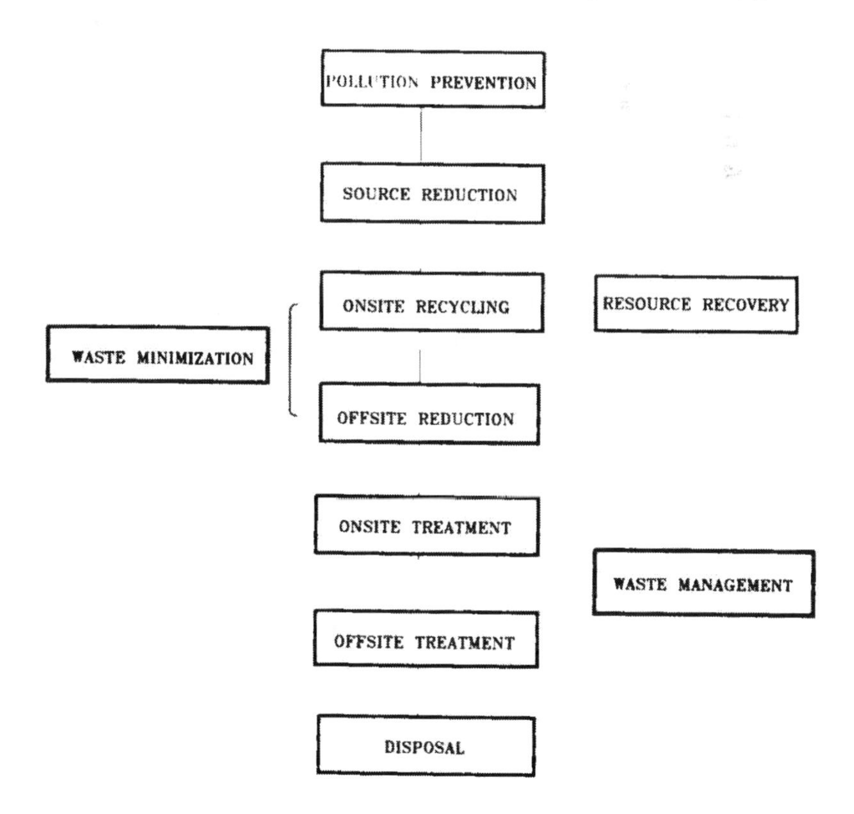

Figure 1-5. Schematic flow-chart showing the hierarchy of waste management options.

reduce the amount of waste generated or via activities performed prior to waste generation that reduce the overall toxicity of the waste. The types of source reduction measures include process modifications, input changes, product substitutions, improved operations, and development of resource recovery programs. Process modifications can incorporate major and minor redesign of manufacturing operations and procedures, including process or equipment changes, or automation, that results in the reduction of waste. Input changes reflect changes in raw materials or feedstocks used in the production process or operation that reduce or eliminate the generation of a waste. Product substitutions or reformulation of products and improvements in waste management practices regarding inventory management, material handling, housekeeping, and education should also be part of conducting business. In addition, reclamation, recovery, and recycling alternatives should be strongly considered and resource recovery programs implemented. Most importantly, all businesses today must develop short- and long-term waste reduction goals and objectives.

Disposal is and rightfully should be demonstrated to be the last option available. Materials that are considered for disposal should only be restricted to those that cannot be reclaimed, reused, or recycled, those that cannot be further treated, those that have been stabilized, solidified, or encapsulated (i.e., hazardous organic wastes), or those materials that are regulated by state and federal land disposal laws.

The second most desirable option is recycling. Recycling is an integral part of today's businesses and industry; it is the meeting place between the economy and the environment. Recycling can generally be defined as the use, reuse, or reclamation of all or part of a hazardous waste, which can occur on- or off-site, and which may be an integral part of a manufacturing process or an independent free-standing operation. Reclamation and use (or reuse) involve two very different technological strategies. Reclamation involves technologies that result in the direct removal of the contaminant from its matrix for the purpose of collection and subsequent reuse. Examples of a reclaimed product are lead recovered from batteries or spent solvents that are regenerated. The soil or material matrix is typically discarded. Use and reuse technologies involve the incorporation of the contaminant and/or its matrix (i.e., contaminated soil) as an ingredient to produce a commercially viable product. Use and reuse do not include the removal of contaminants from a matrix or waste and their collection in a more useful form; nor do they include modifying the contaminants and matrix so that a new product with more desirable properties results. Thus, the main, basic difference between a recycling plant versus a recovery or treatment facility is in the value of the end product. A recycling plant facility, whether on- or off-site, produces an end product that is marketable and can be sold commercially. A recovery or treatment facility, on the other hand, generates a residual that requires further treatment and/or disposal.

As of 1981, very little hazardous waste was recycled. Although 40% of the generators under RCRA reported that they recycled some hazardous waste, only 4% of the total generated was actually recycled. Of this amount, 81% of the hazardous waste recycled occurred on-site, with the remainder being transported off-site to be recycled. Not included in these estimated figures are those wastes that were recycled but exempt from regulation under RCRA. The amount of waste being recycled since 1981 has certainly increased since then, due to increasing concerns regarding long-term liability associated with landfill disposal, land-based restrictions, increased hazardous waste taxes and fees, increasing recognition of certain constituents considered hazardous (i.e., resource recovery), and technology advances and awareness.

A major consideration when formulating remedial or corrective action plans as they relate to contaminated soil is the determination of which technology or combination of technologies will best serve the specific needs of a particular project. No two projects are exactly alike, with each being characterized by its own site-specific parameters and constraints. However, there are four objectives that must always be met if the corrective action is to be successful: the corrective action must be environmentally sound, time efficient, cost effective, and sufficient to alleviate or provide the minimal amount of long-term liability reasonably available, certainly in comparison with other soil remediation options and alternatives. The reuse and recycling of contaminated soil certainly meet these objectives in that reuse and recycling technologies essentially convert materials that otherwise would be disposed of into useful and viable commercial commodities. For reuse and recycling to be successful, the end product must address both environmental and engineering concerns. Environmental concerns include primarily long-term factors such as durability and leachability. Engineering concerns is the need for the end product to serve its intended use in regard to factors such as strength and performance.

Technologies of importance to the reuse and recycling of contaminated soil involve stabilization and solidification. Both of these technologies are closely related and use physical, chemical, and thermal processes to essentially detoxify a particular waste material. Fixation is a term used to refer to either stabilization or solidification. Stabilization does not change the physical nature of the waste; it includes processes that reduce the risk posed by a waste by converting the contaminants into a less soluble, mobile form. Functionally, stabilization can be described in terms of three properties of the final product: mechanical stabilization, immobilization by fixation, and immobilization by isolation. Examples include the ion exchange of heavy metals in an alumina silicate matrix of a cementitous agent or sorption of heavy metals on fly ash in an aqueous system.

Solidification includes processes that encapsulate waste in a monolithic solid of high structural integrity, and durable and stable matrix that is more compatible for reuse, storage, or disposal. Chemical interaction between the waste and the solidifying agent is not a prerequisite, although such interaction may mechanically bind the waste in the monolith. In solidification, contaminant migration is restricted by greatly decreasing the amount of surface area exposed to leaching and/or by isolating the waste within an impervious container. Solidification thus creates barriers between the waste components and the environment by reducing the permeability of the waste or matrix, reducing the effective surface area available for diffusion, or both, with or without chemical fixation.

Encapsulation is another process of importance in implementing reuse and recycling strategies. Encapsulation involves the complete coating or enclosure of a toxic particle or waste agglomerate with a new substance (i.e., an additive or binder such as asphaltic emulsion). Microencapsulation refers to the encapsulation of fine or individual waste particles. Microencapsulation refers to an agglomerate of waste particles or microencapsulated material (i.e., large blocks or containers of waste).

Recycling is essentially a solid-waste management strategy. Compared to landfilling, incineration, etc., recycling is certainly more desirable; it is, in fact, the preferred method of solid-waste management today. Conventional technologies for the recycling and reuse of contaminated soil, their incorporation into a matrix that produces a viable commercial product, are the subject of this book. These technologies include cold-mix asphalt (CMA) processes, hot-mix asphalt (HMA) processes, cement production, and brick-manufacturing processes. Other processes do not necessarily produce a product; thus they are not considered reuse or recycling but rather remediation. Some of the more conventional remediation-type processes include low temperature thermal desorption, incineration, vapor extraction, and bioremediation (Table 1-8). Affected soil is treated and/or disposed of on-site or *in situ* from where it was initially excavated, or at a landfill. Regardless of the remedial strategy being used, subsurface conditions characterized by complex stratigraphy, low permeability soil, fracture permeability, and relatively deep depths and/or aboveground and underground structures will all limit the approach and effectiveness of any remedial technology.

In 1981, only about 4% of the total volume of hazardous waste generated was recycled. Most of the hazardous waste that was recycled, about 81%, was dealt with on-site, with the remainder transported off-site for recycling. These volumes have since increased for a number of reasons, including issues associated with long-term liabilities, land-bans, increased federal taxes on hazardous waste, and increasing recognition of the resource value of constituents and materials once perceived as hazardous waste.

Many states are slowly coming around to more aggressive recycling regulations and programs. State recycling programs are becoming more firmly established, with higher recycling-rate goals being set but rarely met. Although more than half of the states have recycling goals of 40% or more (with 16 at 50% or more), only 18 states reported a recycling rate higher than 25% at the end of 1995. Despite the high expectations, at least ten states anticipate a reduction in their recycling budget in 1996 relative to 1995. Eight states (Arizona, Delaware, Idaho, Kansas, Oklahoma, Utah, Wisconsin, and Wyoming) currently have no recycling or waste reduction goals. The most dramatic reduction is probably Michigan, with a drop from $14 million for their fiscal year 1995–1996 to $400,000 beginning in late 1996, and anticipated elimination altogether for the fiscal year 1997–1998. Such reductions reflect depletion of bond monies, which provided funds to assist with start-up recycling programs, assuming that such programs would eventually be self-sufficient.

In California, as in most states, waste streams are increasing as the population increases. California has implemented an agenda to divert 25% of its solid waste that is disposed of landfills by 1995 and 50% by the year 2000, by initiating a variety of recycling programs. These programs are anticipated to create an additional 20,000 jobs throughout the state. The strategy behind this program is to bring recycling markets to all parts of California by establishing recycling market development zones, providing state and local incentives to persuade manufacturers in these zones to use recycled and secondary products and materials, identifying opportunities for retrofitting manufacturing processes to accommodate the use of recycled and secondary materials and products, and coordinating public and private efforts to promote these material and product markets.

The definition of recycling varies from state to state. Some states consider conventional treatment technologies such as low temperature thermal adsorption and landfarming as recycling technology. Excluded are solidification or treatment techniques that occur directly in or on the land, such as evaporation, surface impoundments, or land-farming techniques. Reuse and recycling in the context of this book include the incorporation of contaminated soil and other materials considered a waste or hazardous waste and the utilization of such material as an ingredient to produce commercially viable products.

1.3 THE CONCEPT OF RESOURCE RECOVERY

The concept of resource recovery is simply not to view contaminated materials or soil as a waste, but rather as a resource of beneficial use and of commercial value. In the course of day-to-day operations, many industrial and manufacturing companies with multiple sites and operations inadvertently generate varying quantities of contaminated soil by the very nature of their business, or have generated contaminated soil and other materials as a result of past operations, activities, and releases. Large releases, such as a breached pipeline or tankage, and emergency spill responses are unavoidably commonplace — if it is mechanical, it will surely break down eventually. In addition, limited quantities of contaminated soil are accidentally generated by certain operations such as valve repairs, pipeline relocations, truck and tankage overfills, routine maintenance, etc. (Figures 1-6 and 1-7). While not always highly visible, such impacts can equate to rather high costs and other liabilities when the overall number of facilities and operations a company is responsible for are taken into account.

Although prevention is the best policy, by their nature, such businesses will continue to generate their fair share of contaminated soil. Development of soil management programs that allow a company to effectively manage its

Figure 1-6. A significant breach of a valve results in the release of product into this bermed but unlined above-ground bulk hydrocarbon storage area.

operations-generated waste in an unobtrusive, cost-effective, time-efficient, environmentally sound manner, while reducing or alleviating long-term liability, will become increasingly important. Resource recovery takes this a step further with the reuse and recycling of such affected soil into commercially viable products that can be utilized on site or on other company-owned sites, thus transforming such material from a waste to a commodity and/or capital assets of true commercial value.

The incorporation of contaminated soil and other materials into asphalt to produce useful, commercially viable products is the sole technology that is applicable to large volumes of affected soil, regardless of whether the soil is contaminated by various petroleum hydrocarbons, organic or inorganic compounds, metals, PCBs, or pesticides and herbicides. Contaminated soil in the construction industry has been and can be considered for use to fulfill a variety of uses and applications, including utilization as aggregate in asphalt, sand substitute, asphalt cement modifiers, asphalt-rubber, chip seal, mineral filler, overlay sealant, joint and crack sealant, and recycled pavement. In addition, affected soil, initially classified as hazardous waste, is incorporated with asphalt emulsion and specific grades of aggregate, utilizing conventional CMA or HMA technology, to produce a range of asphaltic products, which can serve a variety of end uses. Among these end uses are landfill caps and liners, road construction materials, paving, refinery and tank farm dikes and containment

Figure 1-7. Release of hydrocarbon product into an unlined pipeline corridor at an active refinery site.

structures, truck terminal and salvage yard pavements, port facility container shipping yard surfacing, and design fill, among numerous others.

When one views how different types of solid waste are produced vs. how they are applied to recycling, many materials have been recycled, notably, in the construction of highways. These efforts have been summarized by the United States Department of Transportation (USDOT) and have been found to range from use on a very limited experimental basis to wide use and acceptance (Table 1-9). Not indicated as a waste category, however, is the use of contaminated soil, let alone as a recyclable material in the use of paving construction. In regard to current disposal practices by the USDOT, removal

Table 1-9 Summary of Known Recycled Waste Applications

Waste category	Annual rates (millions of metric tons)		Current and past highway uses (millions of tons)				
	Produced	Recycled	Asphalt pavement	Concrete pavement	Base course	Embankment	Other
Agricultural wastes							
Animal manure	1,460		NA[a]	NA	NA	NA	AA
Crop wastes	360		ER[b]	ER	UN[c]	UN	UN
Logging and wood wastes	64		NA	NA	NA	AA[d]	AA
Miscellaneous organics	27		UN	NA	NA	UN	UN
Domestic wastes							
Paper and paperboard	66.7	16.4	NA	NA	NA	NA	LA[e]
Yard waste	31.9	3.8	NA	NA	NA	NA	AA
Plastic[h]	14.7	0.3	LR[f]	ER	UN	UN	AR
Glass	12.0	2.4	LA	NA	LA	LA	LR
Municipal waste ash[h]	7.3		LR	ER	LA	NA	UN
Sewage sludge/ash	7.3		LR	LR	ER	LR	LR
Scrap tires[h]	2.3	0.4	AR[g]	ER	ER	AR	AR
Industrial waste							
Reclaimed asphalt pavement	91	73	AA	LR	AA	AA	UN
Coal fly ash	45	11	AA	AA	AA	LR	UN
Demolition debris	23		ER	UN	ER	ER	UN
Cement and lime kiln dust	21	13	ER	UN	ER	ER	UN
Sulfate waste	16		ER	ER	ER	ER	LR
Coal bottom ash/bottom slag	16	5	LR	UR	LR	ER	UN

Blast furnace slag	14.5	AA	AA	LR	LR	AA
Non-ferrous slags[h]	9	LR	ER	LR	UN	LR
Foundry waste[h]	9	ER	UN	UN	ER	ER
Roofing shingles	6.4	AR	UN	UN	AR	UN
Steel slag	7.3	AR	NA	LL	AA	LA
Reclaimed concrete pavement	3	AR	AR	AA	AA	AA
Lime waste	2	ER	UN	ER	ER	ER
Mining and mineral waste						
Waste rock[h]	930	AA	UN	LR	LR	UN
Mine tailings[h]	473	AR	UN	AR	AL	UN
Coal refuse	109	ER	UN	UN	ER	UN
Phosphogypsum[h]	96	UN	UN	ER	ER	UN
Washery rejects	32	NA	NA	NA	ER	UN

[a] Unacceptable use.

[b] Experimental; design and performance research suggested.

[c] Unknown use.

[d] Accepted use; no further research suggested.

[e] Limited use; no further research suggested.

[f] Limited use; design and performance research suggested.

[g] Accepted use; design and performance research suggested.

[h] There are environmental concerns with these materials that may require further research.

From United States Department of Transportation, 1993, A Study of the Use of Recycled Paving Material, Report Nos. FHWA-RD-93/095 and EPA/600/R-93/095, 34 pp.

Table 1-10 Summary of Disposal Practices

Material/appurtenance type	Average percentage of material (in total volume) Disposed[a]	Reused/recycled[b]
Asphalt concrete: surface course	16	82
Base course	16	82
Stabilized base	27	65
Crushed stone	16	67
Crushed gravel	19	77
Granular subbase	22	73
Stabilized subbase	26	50
Shoulders, asphalt	22	74
Concrete culverts	74	22
Corrugated steel pipe culverts	87	13
Wood culvert	100	0
Multiplate underpass or culvert	66	26
Guard rail	48	52
Guard rail posts (steel and wood)	54	42
Signs — advisory and regulatory	47	53
Sign posts	56	44
Sign or signal pole and structures	54	44
Bridges: aluminum or steel railing	56	44
Steel superstructure and deck	63	37
Concrete beams	83	12
Concrete deck	89	11

[a] These materials may be buried on the project site, landfilled, sold as scrap material, and/or disposed of as contractor's property. These materials may also be reused or recycled.

[b] These materials are functionally reused or recycled in highway projects.

From United States Department of Transportation, 1993, A Study of the Use of Recycled Paving Material, Report Nos. FHWA-RD-93/095 and EPA/600/R-93/095, 34 pp.

or replacement of existing pavement structures produces the greatest quantity of waste material, and recycled asphalt pavement is the most commonly recycled or reused material based on a 1993 USDOT survey (Table 1-10). However, there is no mention of contaminated soil, which is often encountered at construction sites or during property redevelopment, used as a recyclable material.

Contaminated soil can be a necessary ingredient to produce a variety of asphaltic products. Petroleum-contaminated soil has been used as an ingredient of HMA since 1985. The USDOT in a study conducted in 1988 suggested that an estimated 25 million tons of industrial waste could be recycled and consumed by the United States asphalt industry by simply utilizing 5% waste products in all HMA mixes. Approximately 500 million tons of HMA were processed and subsequently applied in 1988. This does not include the estimated 250 million tons of HMA being used for construction, rehabilitation, and maintenance of highways. Consider this in light of overall expenditures for highways in 1988 exceeding $68 billion dollars.

Of an estimated 4 million mi of roads in the United States, about 2.3 million mi of these roads are surfaced with either asphalt or concrete, of which 2.2 million mi or 96% are asphalt-paved. Or of the 2 trillion vehicle mi traveled, about 95% is on asphalt-paved roads. California, for example, with about 57,000 miles of roads, operates one of the largest road systems in the United States. Although concrete is usually used for the main traffic lanes for freeways, asphalt is the pavement of choice for highways, streets, and roads. In Texas, about 13 million forms of coal combustion by-products (CCBP), of which 3 million forms are fluegas desulfurization (FGD) material, are generated each year. Eleven percent of these materials found markets in Texas, compared with 30% nationally. However, the remaining 89% or over 11 million tons of CCBP eventually are disposed of at landfills. As with contaminated soil, use of fly ash, plastics, rubber, and many other materials, which all have a proven track record in road and construction applications, is beginning to be encouraged at both the federal and state levels. Many waste types have been incorporated into a variety of asphaltic end products. The HMA industry has received pressure to incorporate a wide variety of waste material into their mixes. Waste materials incorporated include (1) industrial wastes, such as cellulose wastes, wood lignins, bottom ash, and fly ash, (2) municipal and domestic wastes, such as incinerator residue, scrap rubber, waste glass, and roofing shingles, (3) mining wastes, such as coal mine refuse, mill tailings, and waste rock, and (4) environmental wastes, such as petroleum hydrocarbon-contaminated soil (Table 1-11). Other materials that are now incorporated on a routine basis include blast furnace slags, steel slags, and sulfur.

As of 1993, various state DOT agencies have used a minimum of 16 various waste materials in the production of HMA. Forty-four states have indicated the use of reclaimed asphalt pavement, whereas 38 states have used scrap tires in rubber-asphalt mixes. Fourteen other waste types have also been inducted in HMA, as summarized in Table 1-11. In addition, a minimum of 18 different wastes or by-products have been specified by some states, with the major ones being summarized in Table 1-12. Those most frequently used include reclaimed asphalt pavement, fly ash, scrap tires, blast furnace slag, reclaimed concrete pavement, and steel slag.

The U.S. EPA in 1992 identified approximately 77 permanent HMA facilities in the United States that recycle petroleum-contaminated soil into marketable products, although these facilities are not evenly distributed throughout the United States (Table 1-13). About 22 and 19 facilities are located in EPA's Regions 1 (northeastern United States) and 4 (southeastern United States), respectively. Thirteen facilities are located in Region 5

Table 1-11 Solid Waste Materials and By-Products Potentially Suitable for Asphalt Incorporation

Waste type	Waste description	Annual quantity (millions of metric tons)	DOT State utilization
Domestic household and commercial refuse	Glass	11.0	6
	Plastic	13.0	3
	Incinerator ash	7.8	2
	Scrap tires	2.2	38
	Subtotal	**34**	
Industrial	Coal ash	65.0	9
	Blast furnace slag	15.0	Unknown
	Steel mill slag	7.0	13
	Non-ferrous slag	9.0	1
	Reclaimed asphalt pavement	91.0	44
	Reclaimed concrete pavement	3.0	7
	Foundry wastes	9.0	1
	Roofing shingle waste	7.0	2
	Concrete debris	Unknown	1
	Subtotal	**206**	
Mineral	Waste rock	925	1
	Mill tailings	470	7
	Geothermal scale	Unknown	Unknown
	Subtotal	**1395**	
	Total	**1635**	
Environmental	Contaminated soil	Unknown	Unknown
	Construction debris	Unknown	Unknown

Modified from Kandhal, P. S., 1993, Waste materials in hot mix asphalt — an overview, in *Use of Waste Materials in Hot-Mix Asphalt* (Edited by H. F. Waller), ASTM STP 1193, pp. 3–36. With permission.

(north-central United States), with Regions 3 (east-central United States) and 10 (northwestern United States) having 11 and 5 permitted facilities, respectively. The remaining seven facilities are spread among the other five EPA regions. In their study, the U.S. EPA reported that HMA was the most commonly manufactured product. Petroleum contaminants accepted included gasoline, kerosene, diesel, fuel oil no. 2, fuel oil no. 4, and fuel oil no. 6. In contrast to this study, mobile cold-mix asphalt (CMA) plants, which have the advantage of processing contaminated soil on-site and produce a wider variety of asphaltic end products, are situated nationwide. The actual number of operating CMA plants both stationary and mobile and quantities produced

Table 1-12 Waste Materials and By-Products Incorporated in HMA Via State Specifications

Waste type	Use	Number of states
Reclaimed asphalt pavement	New or recycled asphalt mixes	48
Granulated tire rubber	Asphalt-rubber paving mixtures; stress-absorbing membrane interlayers	22
Fly ash	Asphalt rubber seal coat; mineral filler in asphalt	11
		20
Air-cooled blast furnace slag	Aggregate in asphalt mixes	16
Steel slag	Aggregate in asphalt wearing surface mixes	14
Reclaimed concrete pavement	Aggregate in new asphalt pavement	12
Crushed glass	Fine aggregate in asphalt paving mixes	5

Modified from Kandhal, P. S., 1993, *Use of Waste Materials in Hot-Mix Asphalt* (Edited by H. F. Waller), ASTM STP 1193, pp. 3–36.

Table 1-13 Number of Permitted Recycling Facilities

EPA region	State name	EPA region	State name
1	Connecticut	5	Ohio
1	Maine	5	Wisconsin
1	Massachusetts	6	Arkansas
1	New Hampshire	6	Louisiana
1	Rhode Island	6	New Mexico
1	Vermont	6	Oklahoma
2	New Jersey	6	Texas
2	New York	7	Iowa
3	Delaware	7	Kansas
3	District of Columbia	7	Missouri
3	Maryland	7	Nebraska
3	Pennsylvania	8	Colorado
3	Virginia	8	Montana
3	West Virginia	8	North Dakota
4	Alabama	8	South Dakota
4	Florida	8	Utah
4	Georgia	8	Wyoming
4	Kentucky	9	Arizona
4	Mississippi	9	California
4	North Carolina	9	Hawaii
4	South Carolina	9	Nevada
4	Tennessee	10	Alaska
5	Illinois	10	Idaho
5	Indiana	10	Oregon
5	Michigan	10	Washington

Note: Includes permitted or formally state-approved facilities.

After United States Environmental Protection Agency, 1992, Potential Reuse of Petroleum-Contaminated Soil — A Directory of Permitted Recycling Facilities, USEPA Report No. EPA/600R-92/096, 38 pp.

relative to HMA facilities are inferred to be extensive but are not currently monitored, so actual numbers and quantities generated are presently uncertain.

Although concerns exist that the nation's roads do not become a dumping ground for waste materials in order to simply comply with certain mandates or ease the burden of disposal, identifiable advantages are increasingly being recognized by the regulatory community, industry, and the general public.

BIBLIOGRAPHY

Ahmed, I., 1991, Use of Waste Materials in Highway Construction: Indiana Department of Transportation, Report No. FHWQ/IN/JHRP-91/3.

Andrews, J. S., Jr., 1992, The cleanup of Kuwait, in *Hydrocarbon Contaminated Soils*, Vol. II, (Edited by P. T. Kostecki, et al.), CRC/Lewis, Boca Raton, FL, pp. 43–48.

Blumberg, L. and Gottieb, R., 1989, *War on Waste — Can America Win Its Battle with Garbage?*, Island Press, Washington, D.C., 301 pp.

California Environmental Protection Agency Department of Toxic Substances Control, 1992, Commercial Hazardous Waste Facilities for Recycling/Recovery, Treatment, Disposal: CEPA Draft Document No. 902, 97 pp.

California Assembly Bill No. 1306.

Collins, R. J., 1992, Availability and Uses of Wastes and By-Products in Highway Construction: Transportation Research Board, Transportation Research Record, January, 1992 (reprint).

Conner, J. R., 1990, *Chemical Fixation and Solidification of Hazardous Waste*, Van Nostrand Reinhold, New York, 692 pp.

Czarnecki, R., 1988, Making Use of Contaminated Soils, Civil Engineering, ASCE, December, 1988, pp. 72-74.

Dietz, S. K. and Burns, M. E., 1989, Quantities and sources of hazardous waste, in *Standard Handbook of Hazardous Waste Treatment and Disposal* (Edited by H. M. Freeman), McGraw-Hill, New York.

Electric Power Research Institute, 1993, Chemical and Physical Characteristics of Tar Samples from Selected Manufactured Gas Plant (MGP) Sites: EPRI TR-102184, May, 1993.

Heumann, J. M., Striano, E., and Egan, K., 1996, State Recycling Programs — Ready to Leave the Nest?, *Waste Age*, August, 1996, pp. 35–47.

Kandhal, P. S., 1993, Waste materials in hot mix asphalt — an overview, in Use of Waste Materials in Hot-Mix Asphalt (Edited by H. F. Waller), ASTM STP 1193, pp. 3–36.

Lund, H. F., 1993, *The McGraw Hill Recycling Handbook*, McGraw-Hill, New York.

Means, J. L. et al., 1995, *The Application of Solidification/Stabilization to Waste Materials*, CRC/Lewis, Boca Raton, FL, 334 pp.

PEDCO, 1979, PEDCO Analysis of Echhardt Committee Survey for Chemical Manufacturer's Association, PEDCO Environmental Inc., Washington, D.C.

Saylak, R., 1993, Field Demonstration of the Use of Coal-Fired Power Plant By-Products in Roadway Construction, By-Products Utilization and Recycling, Research Center Project No. 510681-5000, Houston, TX, 13 pp.

Testa, S. M., 1994, *Geological Aspects of Hazardous Waste Management*, CRC/Lewis, Boca Raton, FL, 537 pp.

Testa, S. M. and Winegardner, D. L., 1991, *Restoration of Petroleum-Contaminated Aquifers*, CRC/Lewis, Boca Raton, FL, 269 pp.

United States Bureau of Mines, 1985, Mineral Facts and Problems: United States Bureau of Mines Bulletin No. 671, 1060 pp.

United States Department of Transportation, 1988, Selected Highway Statistics and Charts, Federal Highway Administration.

United States Department of Transportation, 1993, A Study of the Use of Recycled Paving Material, Report Nos. FHWA-RD-93/095 and EPA/600/R-93/095, 34 pp.

United States Environmental Protection Agency, 1985, Report to Congress — Wastes from the Extraction and Beneficiation of Metallic Ores, Phosphate Rock, Asbestos, Overburden from Uranium Mining and Oil Shale, U.S. EPA Office of Solid Waste and Emergency Response, EPA/530-SW-85-003.

United States Environmental Protection Agency, 1992, Potential Reuse of Petroleum-Contaminated Soil — A Directory of Permitted Recycling Facilities, U.S. EPA Report No. EPA/600R-92/096, 38 pp.

United States Environmental Protection Agency, 1992, Characterization of Municipal Solid Waste in the United States, 1992 Update-Executive Summary, Report No. EPA/530-5-92-019, July, 1992.

Wisconsin Department of Natural Resources, 1995, What technologies are used to clean up LUST sites, *Release News*, Vol. 5, No. 1, pp. 12–15.

2 REGULATORY ASPECTS

The Comprehensive Environmental Response Compensation and Liability Act (CERCLA), as amended by the Superfund Reauthorization Act (SARA) provides for federal authorities to respond to releases of hazardous substances, pollutants, or contaminants to air, water, and land at NPL sites. Cleanup standards provided in Section 121 of SARA require that selected remedies be cost-effective and protective of human health and the environment. The federal cleanup standards of SARA also encourage highly reliable remedial actions that provide long-term protection. This mandate includes permanently and significantly reducing the volume of materials considered hazardous and toxic, including contaminated soil. This reduction can be accomplished by one or a combination of several technologies and regulatory strategies (within the regulatory framework).

2.1 INTRODUCTION

Whether waste or material is considered hazardous depends on the nature and concentration of the toxic constituents contained within the waste material, their mobility and persistence in the environment, the quantities of the waste generated, and the potential for mismanagement. It is difficult to truly determine exactly what a waste is. The mere fact that a material may contain a specific constituent at an elevated concentration or fail a certain test or series of tests does not in itself make the affected material a waste. Only when the material is deemed capable of posing a substantial present or potential hazard to human health or to the environment when improperly treated, stored, transported, disposed, or otherwise managed, does it then become a candidate for consideration as a waste. It has thus become increasingly important that prior to formulation and implementation of remedial and recycling strategies, a thorough understanding of the pertinent federal, state, and local regulations is achieved.

As will be discussed in this chapter, the regulatory posture over the past several years has been to redefine the spirit, letter, and intent of environmental regulations concerning perceived hazardous and toxic wastes. Presented is a discussion of the responsibility of the generator, what characterizes a waste, the regulatory framework that allows for preferred reuse and recycling, waste declassification criteria, exemptions, and transportation considerations.

2.2 GENERATOR RESPONSIBILITY

A generator can be any person or facility who, by ownership, management, or control, is responsible for the creation of hazardous waste. The generator of any material potentially considered a waste or hazardous waste maintains several important responsibilities. The generator must first determine whether a particular waste is excluded from regulation. Should the waste not be excluded via any means, then the generator is responsible for hazardous waste determination and identification. If a waste is determined to be hazardous, then the generator has the responsibility to identify and implement, as appropriate, possible exclusions or restrictions pertaining to management of the specific waste. A generator is also responsible for pretransport requirements, such as labeling, reporting, and record-keeping requirements, manifesting requirements, proper treatment or disposal of material, and personnel training. A generator's responsibility to self-classify applies to any waste, including contaminated soil. In addition, a generator is also exposed to present and future liabilities under CERCLA for any waste it produces.

As part of any program for recycling of contaminated soil, the generator may also have other responsibilities based on state or local regulations and guidelines, including the proper management of the program. The generator and user may be required to demonstrate that the contaminated soil will provide beneficial qualities to the end product, or adhere to all regulatory and industry standards pertinent to the nature of the end product (i.e., ASTM, USDOT, etc.), to ensure that the product will be viewed as contributing to the quality of the product.

2.3 WASTE CHARACTERIZATION

Under federal law, the definition of a hazardous waste can be found under certain acts within the Federal Code of Regulations (CFR). These acts include the Resource, Conservation and Recovery Act of 1976 (RCRA), the Comprehensive Environmental Response, Compensation, and Liability Act of 1980 (CERCLA), and Superfund Amendments and Reauthorization Act of 1986 (SARA). These regulations include lists of waste known to exhibit certain toxicological properties with respect to human health and the environment. RCRA is the primary federal legislation governing hazardous waste activities, including generation, transport, treatment, storage, and disposal of hazardous waste. Under CERCLA and SARA, the USEPA is responsible for determining the methods and criteria for removal of waste and residual contamination from

Table 2-1 Summary of Primary Federal Regulations Regarding Recycling of Contaminated Soil

40 CFR Section (part)	Subpart	Description
260.10	A	Definitions
261.2		Definition of a solid waste
		Used in a manner constituting disposal
261.3		Definition of a hazardous waste
261.4		Exclusions
261.6		Requirements for recyclable materials
261.10	B	Criteria for identifying the characteristics of hazardous waste
261.11		Criteria for listing hazardous waste
266.20	C	Recyclable materials used in a manner constituting disposal

a particular site. SARA strongly recommends remedial actions by treatment that permanently and significantly reduce the volume, toxicity, or mobility of hazardous substances. In addition, CERCLA remedial actions also require overall protection of human health and the environment, compliance with Applicable or Relevant and Appropriate Requirements (ARARs), short- and long-term effectiveness and permanence, implementation feasibility, and state and community acceptance. A summary of pertinent federal regulations pertaining to the recycling of contaminated soil is presented in Table 2-1.

2.3.1 Solid Waste Characterization

From a regulatory perspective, it is the responsibility of the generator to determine whether a particular material is to be considered either a hazardous or toxic waste. In order for a material, including contaminated soil, to be considered a hazardous or toxic waste, it must first be determined that it is a solid waste. A solid waste can best be described as anything that is discarded or abandoned (i.e., disposed of, burned or incinerated, or accumulated, stored or treated, but not recycled, before or in lieu of being abandoned by being disposed of, burned, or incinerated). Upon demonstrating that the material is a solid waste, the determination is made as to whether it can be proven hazardous or toxic.

A material is not a solid waste when shown to have been recycled (40 CFR 261.2 (e)), which is characterized as materials:

1. Used or reused as ingredients in an industrial process to make a product, providing the materials are not being reclaimed;
2. Used or reused as effective substitutes for commercial products; or

3. Returned to the original process from which they are generated, without first being reclaimed. The material must be returned as a substitute for raw material feedstock, and the process must use raw materials as principal feedstocks.

There are several materials that are not considered a solid waste. These materials are referred to as exclusions and include (1) mining overburden that has been returned to the mine site, (2) drilling fluids, produced waters, and other wastes associated with the exploration, development, or production of crude oil, natural gas, or geothermal energy, and (3) fly ash waste, bottom ash waste, slag waste, and flue gas emission control waste, generated primarily from the combustion of coal or other fossil fuels, excluding facilities that burn or process hazardous waste, among others (40 CFR Part 261.4 (b)(B).

2.4 REGULATORY FRAMEWORK

The regulatory framework of the federal environmental laws and that of many states do not deem everything as "hazardous" and mandate its disposal in a landfill or by incineration. In fact, a review of current regulations proves quite the contrary. The letter, spirit, and intent of current hazardous materials legislation is to promote and develop alternative technology that encourages the use, reuse, and recycling of materials rather than the archaic load, haul, and dump remediation techniques that have produced more environmental problems than they ever solved.

The reuse and recycling of contaminated soil is addressed under the following enabling federal legislation:

- Code of Federal Regulations (CFR) Title 40, Part 261, Section 2 (40 CFR 261.2), Definition of Solid Waste
- 40 CFR 261.2(e), Materials that Are Not Solid Waste When Recycled

The specific objectives of any reuse and recycling strategy are

- To effectively reuse affected soil as an ingredient in a stable, non-hazardous product that would be beneficially used;
- To reduce generator liability to a minimum by complying with pertinent federal and state regulations;
- To reduce the cost of remediation by reusing affected soil as an ingredient in a viable commercial product, thereby eliminating many of the hazardous waste taxes and pretreatment and landfill disposal costs;

- To demonstrate that the reuse and recycling method to be used effectively stabilizes the hazardous constituents comprising affected soil;
- To demonstrate that the reuse and recycling method to be used is a cost-effective, time-efficient, and environmentally sound alternative to landfill disposal of hazardous waste.

2.5 DECLASSIFICATION

Declassification protocols are pursued to reduce the overall volume of contaminated soil and other materials being sent to landfills for disposal and to minimize the costs associated with legal, administrative, and remedial programs. Contrary to initial impressions, the intent of the United States hazardous waste regulations is not to list all contaminated soil and water as hazardous or toxic, but rather to allow avenues for declassification of the material being classified.

A variety of materials initially or at first glance considered hazardous or toxic can be declassified and documented as neither hazardous nor a waste via use of the appropriate federal regulations pertaining to waste reduction and waste minimization. The crux of declassification is the determination of whether a material is a solid or hazardous waste. Before the material can be considered a hazardous waste, it must first be demonstrated that it is a solid waste (i.e., a discarded material) as presented under 40 CFR 261.2(a)(1). The material cannot be classified as a hazardous waste if it is not a solid waste. In becomes very clear upon review of the federal regulations that materials are not considered solid waste if it can be shown that they can be recycled, such that they are

- Used or reused as ingredients in an industrial process to make a product, provided the materials are not being reclaimed
- Used or reused as effective substitutes for commercial products

The responsibility of the generator to understand and properly declassify their materials is of utmost importance in avoiding the stigma of being a generator of hazardous and toxic waste.

Some states go further. In California, recyclable material that is or will be (1) used or reused as an ingredient in an industrial process to make a product or (2) used or reused as a safe and effective substitute for commercial products is excluded from classification as a waste. Hence, if the regulations do not classify recyclable materials as "waste" and these materials are not regulated as "hazardous waste," their use, reuse, and recycling are within the letter, spirit, and intent of environmental legislation.

2.6 EXEMPTIONS

Due to the circuitous nature of the federal regulations, exemptions exist. One of the most misunderstood and often misinterpreted exemptions is the use of recycled materials in a manner such that they are perceived as being "used in a manner constituting disposal" as presented under 40 CFR Part 261.2(c)(1). "Used in a manner constituting disposal" is defined as any types of recycled materials that are applied to or placed on the land or used to produce products that are applied to or placed on the land. RCRA-regulated recyclable materials (i.e., RCRA hazardous waste) must generally be managed as a hazardous waste if they are to be used in a manner constituting disposal. In other words, all requirements pertinent to the generation, handling, and transportation of these materials must be complied with. Facilities receiving contaminated soil characterized as RCRA hazardous waste must also maintain the appropriate hazardous waste permits. Products incorporating contaminated soil to produce road pavement, building foundations, liners and rip rap, and building blocks used in walls are examples of use constituting disposal. Obviously this does not mean that should the material be recycled or reused, one cannot do anything with it, thus discouraging waste reduction, waste minimization, and beneficial uses. A viable, commercial product must be produced, which provides no significant hazard to the recycling operations or the final end product.

Another exemption to this exemption also exists, referred to as "Recyclable Materials Used in a Manner Constituting Disposal" as presented under 40 CFR Part 266, Subpart C, Section 266.20(b). Stated is that products produced for the general public's use that are used in a manner that constitutes disposal and that contain recyclable materials are not presently subject to regulation if the recyclable materials have undergone a chemical reaction in the course of producing the products so as to become inseparable by physical means, and if such products meet the applicable treatment standards as presented under 40 CFR, Subpart D, Part 268, or applicable prohibition levels as presented under Section 268.32 or RCRA 3004d. For the end product not to be construed as "use constituting disposal," the following criteria must be met:

- The recyclable material must be mixed with other material;
- A demonstration must be performed that the elevated concentrations of hazardous constituents of concern as identified in the soil do not exceed such levels in the end product (i.e., leachability testing);
- No significant hazard to the manufacturing process or the product is demonstrated;
- The product needs to be made available to the public, although the product can also be used on-site.

Some states such as California take a less restrictive approach and have further excluded certain materials (non-RCRA regulated waste or waste not regulated by the USEPA, but regulated by the State of California) from the definition of use constituting disposal. Such materials that are considered use constituting disposal are non-RCRA hazardous waste used in the production of asphalt and concrete. Also included are materials that are mixed with other materials and have become chemically bound or physically encapsulated in the product and pass appropriate leachability tests (i.e., do not exceed established levels for STLC or TTLCs, TCLP, etc.). Although also restricted to unused petroleum hydrocarbons, hazardous inorganic or organic constituents can be exempted and subsequently recycled if such constituents do not comprise more than 10% and 5% by weight of the total volume, respectively. Elevated amounts of asbestos, beryllium, cadmium, mercury, and selenium are also excluded.

Local agencies may also maintain certain exclusions. For example, the contaminated soil should not be treated prior to its use in the manufacturing process. In addition, record-keeping, labeling, storage, and business plan requirements may also need to be observed. For example, a business license must be held by the recycler, a seller's permit may be required to sell the product, and the recycler may be required to show operating records, sales receipts, etc., pertaining to the disposition of the product. The recycler must also keep records of the testing of the product to ensure that the appropriate specifications (i.e., structural integrity) are met.

2.7 TRANSPORTATION CONSIDERATIONS

The generator of the recyclable material may be requested to notify the lead regulatory agency prior to the use or transfer of the product for use. Notification may include:

- Generator's name, address, telephone number, and contact person;
- Brief narrative description of the recyclable material (i.e., contaminated soil, etc.);
- Characteristics and concentrations of hazardous properties of the recyclable material;
- Estimated weight of the recyclable material;
- Intended weight of the recyclable material;
- Name, address, and telephone number of the designation accepting the recyclable material.

In most states, only nonhazardous material can be transported off-site to a soil recycling facility that is not considered a USEPA treatment, storage, or disposal facility. This restriction reflects constraints imposed by Department of Transportation regulations and may also reflect several local agencies' restrictions pertaining to air quality, water quality, and other concerns. This restriction does not apply, of course, when the receiving facility is permitted to handle hazardous waste. When transportation is required for moving the contaminated soil from the site of origin to the permanent recycling facility, a Bill of Lading is used. The Bill of Lading is annotated with the acceptance number, and copies are retained by the recycling facility representative for verification of each load. One copy of the Bill of Lading for each load is filed at the facility, along with a copy of the weight record for that load and analytical data corresponding to the acceptance number. Transportation can be performed by semi-end dump trucks or rail cars if convenient. Backup records of each acceptance number, complete with Bills of Lading, weight receipts, and analytical data are routinely kept at the recycling facility or place of business.

Handling nonhazardous or hazardous soil and subsequent processing of such soil and other material on-site present a completely different set of circumstances. Under this situation, both categories of soil can be processed in the production of CMA or HMA, regardless of their quantities or concentrations, provided that processed material is made available to the public as a commercial product. The end product can still be used on-site, as long as it has been made available to the public.

BIBLIOGRAPHY

Federal Code of Regulations, Title 40, Environment and Health, Parts 260, 261, and 266.

Preston, R. L. and Testa, S. M., 1991, Permanent fixation of petroleum-contaminated soils, in Proceedings of the National Research and Development Conference on the Control of Hazardous Materials, Hazardous Materials Control Research Institute, pp. 4–10.

3 SOIL REUSE AND RECYCLING TECHNOLOGIES

3.1 INTRODUCTION

The technologies available for handling contaminated soil vary depending on whether such material is being remediated or treated, reclaimed, used, reused, or recycled. Contaminated soil can undergo a particular remediation technology or combination of technologies prior to being reused or recycled. Remediation oriented technologies essentially reduce the amount of contaminants to levels acceptable for handling, prior to further treatment, disposal, or closure, assuming that acceptable regulatory levels have been reached or negotiated. In many cases, acceptable levels are never reached due to physical, technical, or economic limitations. Recycling differs from a remedial approach in that it involves the incorporation of the affected soil or material via some process to produce a commercially viable product, which fulfills certain environmental and engineering criteria.

Marketable products that can be produced from contaminated soil and other materials (i.e., slag, fly ash, rubber, plastic, etc.) can be essentially divided into two general groups, bituminous- and nonbituminous-containing materials (Table 3-1). Bituminous-containing materials contain aliphatic, mononuclear aromatic, and polynuclear aromatic hydrocarbons. With bituminous-containing materials, affected soils can be processed by one of two technologies, cold-mix asphalt (CMA) and hot-mix asphalt (HMA) technologies. Nonbituminous materials include clay, shale, and other materials. With nonbituminous materials, affected soils can be processed by utilizing cement and brick-manufacturing technologies. Presented in this chapter is discussion of these technologies which all have the ability to incorporate contaminated soil and other materials into viable commercial products. A discussion of reclaim technology is also presented.

Table 3-1 Summary of Reuse and Recycling Technologies

Technology	Process	Media	Contaminant	Possible end use	Advantages	Limitations
Asphalt incorporation	Cold-mix asphalt	Fly ash Foundry sand Grit Mine tailings Plastic Rubber Sand blasting grit Sediment Slag Sludge Soil	Inorganics Low level radioactive waste Metals Pesticides and herbicides Petroleum hydrocarbons PCBs PNAs and MAHs LLW and ILW	Construction material Containment Design fill Insulation Pavement construction Liner	Variety of contaminants acceptable Can incorporate large volumes of affected soil Flexible mix design and specifications Mobile Minimal weather restrictions Cost effective Can be stockpiled and used when needed Technology in place Can accommodate fine-grained, low permeability soil Processing can occur on-site	Volatiles require control Small volumes of contaminated soil may not be economically viable for mobile plants
	Hot-mix asphalt	Sediment Sludge Soil	Petroleum hydrocarbons PNAs and MAHs LLW and ILW	Construction material Containment Insulation	Technology in place Minimal excessive capital costs required to set-up Can process small volumes of affected soil easily Accessible in many areas	Requires application immediately after processing Restricted mix design and specifications Potential elevated emissions could affect plant efficiency Incomplete burning of light-end hydrocarbons can affect quality of end product Emission restrictions

Cement production	Wet	Soil Fly ash Slag Foundry sand	Petroleum hydrocarbons Metals	Cement products Construction materials	Can accommodate wide variety of contaminants and material Technology in place Raw materials readily available Relatively low water solubility Relatively low water permeability	Material restrictions both technically and aesthetically Alkali limitations Odorous material limitations Processor typically interested in raw materials, not energy recovery Wide range of volume increase
	Dry	Soil Metals Fly ash Slag Foundry sand	Petroleum hydrocarbons Metals			
	Dry with preheating and/or precalcining	Soil Fly ash Slag Foundry sand	Petroleum hydrocarbons Metals			
Brick manufacturing		Soil Fly ash	Petroleum hydrocarbons Metals	Abrasives Architectural dimension stone Bricks Castable refractories Ceramic products Cermaic tile Insulation Kiln-lining bricks Piping Roof tiles	Can accommodate fine-grained, low permeability soils Technology in place Processing can occur on-site	Restricted primarily to petroleum hydrocarbons and fly ash

3.2 REUSE AND RECYCLING TECHNOLOGIES

3.2.1 Asphalt Incorporation

The conversion of asphalt into concrete (i.e., asphalt concrete, bituminous concrete, etc.) involves producing a material that is plastic when being worked and sets up to a specified hardness sufficient for its end use. The incorporation of contaminated soil into bituminous end products is accomplished by two conventional processes, CMA and HMA.

3.2.1.1 Cold-Mix Asphalt Processes (CMA)

Commonly referred to as environmentally processed asphalt, the CMA process utilizes contaminated soil, which is essentially used as an ingredient in the production of CMA. CMA is comprised of three basic components, soil, aggregate, and water-based emulsion. Contaminated soil, formerly considered or classified as hazardous waste, is incorporated with asphalt emulsion and specific grades of aggregate to produce a wide range of cold-mix asphaltic products. The contaminated soil serves as the fine-grained component of the asphaltic product. A fourth ingredient occasionally added in small amounts to enhance stability of the end product usually consists of lime, Portland cement, or fly ash.

The incorporation of contaminated soil into a variety of cold-mix asphaltic products can be physically accomplished by two means, mixed-in-place methods for larger quantities and windrowing for relatively smaller quantities. Mixed-in-place processing utilizes a portable asphalt batch plant and can take place at a permanent site or facility or can be considered mobile and set up at a convenient site where contaminated soil awaits processing, regardless of where the end product is to be utilized.

The plant operation consists of a mechanical screening plant (including rolling stacks which are optional), a transfer conveyor, an electric generator, the asphalt plant or pug mill, and an asphalt emulsion truck (typically 5000 gal). All equipment except the rolling stack is electrically powered. Production rates typically range from 100 to 150 tons of CMA per hour.

Material or contaminated soil is introduced into the process through the screening unit or grizzly, which serves to separate all deleterious materials (i.e., trash, plastic, large rocks, etc.) from the soil, and to size material according to design criteria (Figure 3-1). For example, asphalt subbase would use 1½-in. aggregate, whereas pavement may call for ⅝-in. or less. From the screen, the materials travel on a transfer conveyor to the batch plant's soil hopper (Figure 3-2). There are two soil hoppers on the end of the batch plant, one for contaminated soil and the other for aggregate. These materials are fed at a predetermined rate from each hopper using variable-speed conveyors and adjustable feed gates. The feed hoppers discharge onto the batch plant's transfer conveyor to the mixing chamber. The materials discharge from the conveyor into a fluffer wheel compartment where further mixing occurs. Inside this compartment the required amount of emulsion is applied via an asphalt spray bar.

Figure 3-1. Core of CMA showing two different mix designs.

Figure 3-2. Typical screening unit or grizzly utilized for separation of oversized deleterious materials.

From the asphalt compartment, the material discharges into the mixing chamber. Inside this chamber are two counter-rotating paddle-wheel mixers

Figure 3-3. Mobile CMA batch plant showing the two soil hoppers (right) and output of produced CMA (left). Emulsion truck is situated behind batch plant.

that have adjustable rotating speeds. This provides the proper retention time to ensure a complete blending of all materials. The product produced can be utilized immediately or stockpiled for months until needed (Figure 3-3). Prior to placement, however, the product is cured by allowing water to evaporate from the water-based emulsion. What water remains is essentially driven out during compaction (Figure 3-4). Product specifications are typically in accordance with those used by county and state road departments and can be formulated to exceed such specifications to produce road-mixed asphalt. During curing, moderate to high ambient temperatures with low moisture are preferred.

Portable CMA set-ups include

- Portable CMA pugmill,
- Vibratory screening plant (and grizzly),
- Sixty-foot transport conveyors (one extending from the screen to the asphalt plant and one from the asphalt plant to the stockpiles),
- Ninety-foot radial stack conveyor (optional),
- Loaders (two 950-Cat or equal, preferably),
- Generator (150 kW),
- Asphalt emulsion truck (to be shuttled on- and off-site),
- Water truck.

Figure 3-4. Stockpiles of CMA alongside freeway in southern California to be used at a later time.

Operations personnel, depending on the size of the project, can range up to eight, including

- One site supervisor and/or equipment operator,
- Two screen/asphalt plant operators,
- Two equipment/water-truck operators,
- One asphalt emulsion truck operator,
- Two laborers.

Windrowing is applicable for smaller quantities. This process involves coating the soil with an emulsion and simply mixing the materials in place (Figure 3-5). The windrow is typically flattened, with the asphalt applied from a distributor. The asphalt is applied uniformly upon the layer of soil or aggregate at typical rates of 2.3 to 4.5 l/m^2 (0.50 to 1.0 gal/yd^2) at a specified temperature. The asphalt application is then blade-mixed in quantities not usually exceeding 4.5 l/m^2 (1.0 gal/yd^2) per application. Following the first application of asphalt, the soil and/or aggregate are thoroughly mixed by motor graders. Mixing continues until the asphalt is uniformly distributed over the soil and/or aggregate. The mixed material is then again windrowed or spread over the area being surfaced to a uniform depth. This process is repeated until the total amount of asphalt emulsion specified is mixed with the soil/aggregate, and a thoroughly uniform mixture has been obtained.

In the blending of imported aggregate of contaminated soil, aggregate, and asphalt emulsion via windrowing, the windrow needs to be carefully

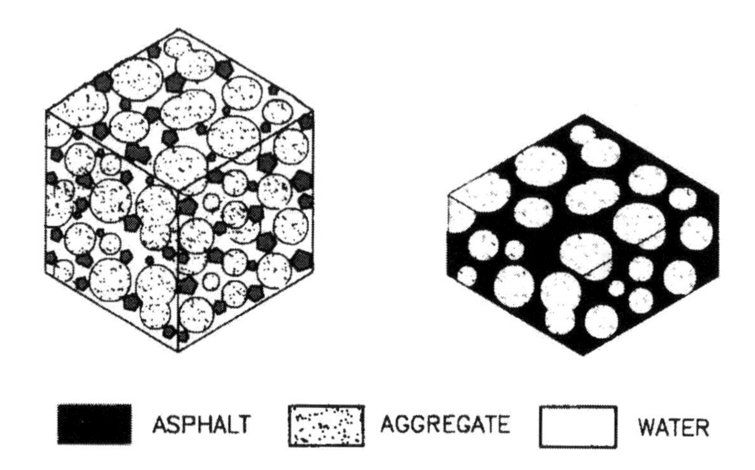

Figure 3-5. Schematic illustrating pre- and post-compaction conditions for CMA.

calculated and maintained. This can be accomplished by thoroughly mixing the soil and aggregate prior to adding the emulsified asphalt. The volume of aggregate in the windrow can be determined as follows (Figure 3-6):

$$Va = \frac{(A + B)C}{2} \times \text{meters (feet)}$$

where Va = volume of aggregate in windrow $m^3/m(ft^3/ft)$, and A, B, C = dimension of windrow, m (ft). The application rate can be determined as follows:

$$Ab = \frac{Va \times Wa \times Pb}{Wb}$$

where Ab = application rate of emulsified asphalt [l/m(gal/ft)], Va = volume of aggregate in windrow, Wa = loose unit weight of dry aggregate, $kg/m^3(lb/ft^3)$ (refer to ASTM Test Method C 29, or AASHTO T 19), Pb = design percent of emulsified asphalt by dry weight of aggregate in the mixture expressed as a decimal, and Wb = weight of emulsified asphalt, [kg/l(lb/gal) = 1.00(kg/l) (or U.S. Customary 8.3lb/gal). The above equations are approximate and based on emulsion at a temperature of 15.6°C (60°F).

The suitability of the resulting asphalt end product is reflected in certain properties of the product once laid down. These properties include stability, durability, flexibility, skid resistance, permeability, and workability. Stability is the asphalt's resistance to deformation. Durability is its resistance to weathering, crushing, and degradation. Flexibility is the product's ability to conform

Figure 3-6. Technique of windrowing for mixing smaller quantities of contaminated soil, emulsion, and aggregate.

to long-term variations within the base due to settling and the ability to bend without fracturing. CMA weathers distortion due to frost or sub-grade weaknesses better than more rigid-type pavements. Skid resistance allows reasonable traction under varied weather conditions. CMA tends to be self-healing under solar heat and rubber-tired traffic. To resist weathering, low permeability is desired and characteristic of CMA. Also important is workability, which is the ability to achieve a smooth finish once the pavement is laid down and compacted. CMA properly applied can provide a smooth surface as well as a high coefficient of friction.

From a technical perspective, a variety of contaminants can be incorporated into CMA, including petroleum hydrocarbons, aromatic and semi-aromatic hydrocarbons, PCBs, pesticides and herbicides, and metals, among others, as discussed in more detail in later chapters. An advantage over HMA is that varied mix designs can be formulated based on the end use of the product; thus the volume of contaminated soil that is incorporated can vary greatly. Since specifications for CMA also vary greatly in comparison to the more rigid specifications for HMA, a wide range of end products can be produced.

There are numerous other benefits associated with CMA. CMA plants can be very easily transported and set up quickly. A wide range of asphaltic end products can be produced (as is further discussed under Chapter 11). Mix designs are quite flexible and can be adjusted to meet specific requirements. Mixtures can be adjusted to complement the type of soil being utilized and are readily receptive to varied aggregate (crushed concrete and asphalt, plastic, slag, etc.) and contaminant types (organics and inorganics, and other materials such as fly ash, plastic, slag, etc.). The product can be stockpiled for several months and used as needed, with no significant adverse or deleterious effects. Since CMA plants are mobile and can be used on an as-needed basis, a continuous waste stream is not required, although on-site CMA processing may not be cost-effective should only a one-time, small quantity occurrence exist. Fume emissions are virtually absent. Since the emulsion used is water based, potential fire and safety hazards are reduced due to the absence of a dryer and heated or high temperature mixes (i.e., heated aggregate). Once placed, pavements can be opened to traffic within 24 to 48 h. CMA is also produced at a fraction of the cost of HMA.

3.2.1.2 Hot-Mix Asphalt Processes (HMA)

Hot-mix asphalt (HMA) technology is most often applied to pavement. HMA is conventionally produced by one of two processes, batch and drum dryer mixers. This technology essentially consists of both mixing and heating to produce pavement material (Figure 3-7). Mineral aggregate, such as crushed rock, gravel, and sand, is blended and dried by heating at temperatures ranging from approximately 300 to 350°F. Hot asphalt is mixed with the heated aggregate at about 5 to 10% by weight to produce pavement material. Aggregate and contaminated soils are typically independently stored in cold bins. Metered amounts of aggregate and contaminated soils are conveyed from the cold feed bins by use of a cold elevator (or belt feeder) to a collecting conveyor, and then enter at the end opposite the burner. The dryer operates at temperatures ranging from 500 to 800°F. The mixture departs from the dryer at a temperature of about 300°F. A hot elevator (or bucket elevator) then conveys the material to a set of screening units for size separation and subsequent storage in hot bins according to aggregate size. The mix design to be used

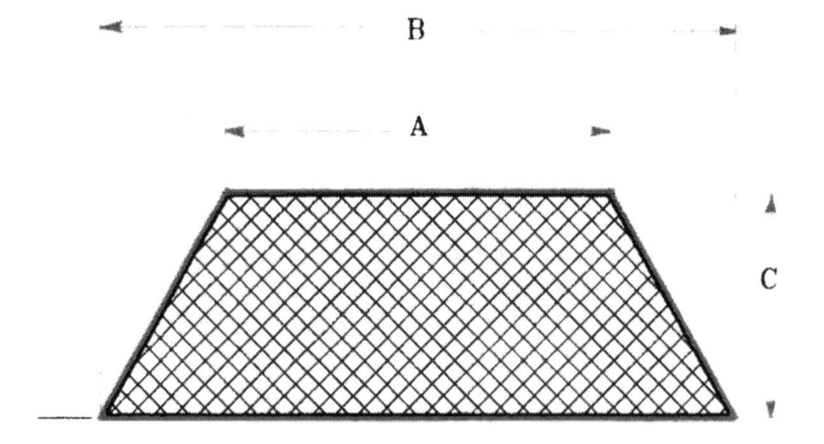

Figure 3-7. Dimensions in determining volume of a windrow.

will determine the measured amount of each size fraction to be used, which is weighed and then dropped into the mixing unit containing hot asphalt. After mixing, the asphalt is conveyed to heated storage containers or to trucks for immediate use.

Earlier attempts to process contaminated soil involved incorporating the affected soil with virgin aggregate, followed by processing in a normal manner. However, elevated hydrocarbon levels in the stack resulted. Test runs performed in California on such systems showed about a 65% reduction in the destruction of hydrocarbons. The reason for this is illustrated in Figure 3-8, which compares dryer length with temperature. Hot gases enter at one end at 2000°F, and cold, wet aggregate enters at the other end. The wet aggregate immediately reaches 212°F, at which time hydrocarbon and water evaporation occurs. Once all the liquids are evaporated into the exiting gas stream, then the temperature starts to rise, heating the aggregate until it reaches 350°F. At this temperature, the aggregate leaves the dryer and enters the bucket elevator. The hot gases enter at about 2000°F and leave the dryer at 300 to 350°F. Since the evaporated hydrocarbon is entering a low temperature gas stream at 300 to 400°F, very little incineration takes place. Certain HMA plant operations have thus modified their process with an incineration-type unit in lieu of an evaporation-type unit.

With an incineration-type unit, vaporized hydrocarbon is incinerated and destroyed, as it is carried through a 5-ft long, 5-ft diameter ceramic cylinder filled with a flame that extends beyond the cylinder for about 11 ft into the dryer, while rotating counter to the rotation of the dryer at 0 to 6 rpm; holdup time of the soil in the ceramic cylinder is at least one minute. Water and hydrocarbon are immediately flushed off as the soil enters into the 2000°F flame within the rotating cylinder. The soil is carried out of the cylinder at about 800°F. Exposure to the 2000°F flame is calculated to be 1.2 sec, based on a 16-ft flame length and a gas velocity of 800 ft/min (i.e., 16 ft per 800 ft/min × 60 sec/min = 1.2 sec). Incorporating a ratio of about 95% virgin to 5 to 10% contaminated soil, an acceptable end product can still be produced.

When incorporating petroleum-contaminated soil, the hydrocarbon contaminants are not destroyed during the drying process but rather during the aggregate preparation process. Volatilization and low-temperature thermal destruction of the organic compounds occur in the dryer, whereas the heavier hydrocarbon contaminants are incorporated into the asphalt-aggregate mix and are subsequently available for use as pavement. Exhaust treatments utilizing cloth filters or baghouses are used to control particulate emissions.

A typical HMA mix contains about 50% coarse aggregate or gravel, 40% fine aggregate or sand, 5% mineral fill such as crushed stone dust or lime, and 5% asphalt cement. Relative grain sizes range from 1.5 in. to No. 4 sieve, No. 4 sieve to less than No. 200 sieve, and less than No. 200 sieve, respectively. The asphalt cement (normally HFMS – 2h) typically contains high concentrations of both mononuclear and polynuclear aromatic hydrocarbons, nitrogen, sulfur, oxygen, and trace amounts of metals or organic metallic compounds.

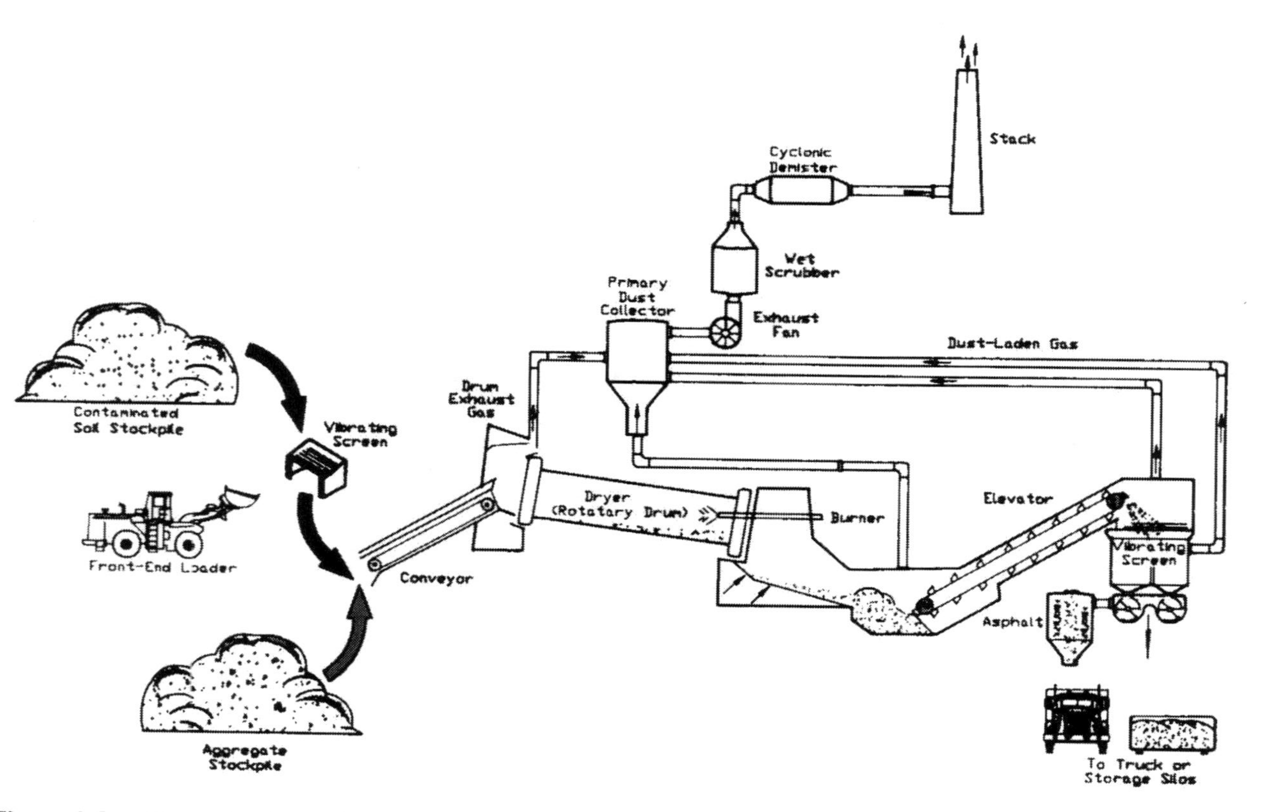

Figure 3-8. General schematic of a HMA bituminization plant.

When incorporating petroleum-contaminated soil, relatively small quantities are utilized and added to the aggregate feed to maintain product quality and minimize air emissions resulting from volatilization of organics in the dryer. HMA plants typically limit the amount of clay and silt content in soil feed to 15 to 20%, although some increase this amount to produce road bedding aggregate and daily landfill cover in addition to asphalt. Some facilities produce this intermediate product for road base or an asphalt component.

HMA production is attractive in that it involves incineration technology, which is easily available and in-place at many facilities; thus, no excessive capital costs are required and only few modifications are deemed necessary (i.e., on the order of $10,000 or more). Small quantities of soil can be processed economically. Large quantities or continuous runs are not necessary. With over 5000 HMA plants throughout the United States, accessibility is fairly good. In addition, no continuing liability for the generation exists.

HMA production incorporating contaminated soil is dependent, however, on the physical and chemical characteristics of the affected soil. HMA design is also restricted in order to maintain required strength and durability according to very stringent specifications and requirements, and is largely restricted according to aggregate type, particle size and volume, soil volume, and clay content. Contaminated soil to be used must be compatible with the HMA mix requirements. This usually limits the amount of fines to 2 to 10% in total volume. High contents of any minus 200 mesh material, 20% or more, may cause deterioration in quality of the end product and contribute to potential stripping problems. An average HMA content comprises 60% coarse aggregate, 40% fine aggregate (i.e., sand and affected soil), and 5.5% liquid asphalt. The total amount of minus 200 mesh material deemed acceptable in the end product is approximately 6%.

HMA plant efficiency may also be jeopardized with an increase in emissions for a feed mixture of clean aggregate and contaminated soil, with volatile emissions from diesel fuel- and gasoline-affected soil by 20 lb/h to 64 lb/h and 67 lb/h, respectively. HMA plants are stationary fixtures and normally do not operate during cold weather. The end product must also be used immediately upon processing. Furthermore, lighter-end hydrocarbons not burned off completely may potentially adversely affect the end product by softening the final mix and affecting curing time.

Estimated cost for incorporating petroleum-contaminated soil into HMA ranges from $40 to $100 per ton, excluding consulting, transportation, and storage costs, averaging about $80/yd^3. Retrofitting a HMA plant for reuse and recycling is estimated at $10,000 to $100,000 for such capital costs as soil storage, feed, conveying, and metering systems.

3.2.2 Cement Production Processes

Cement manufacturing incorporates raw materials such as limestone, clay, and sand, which are typically fed into a rotary kiln. Introduced into the raised

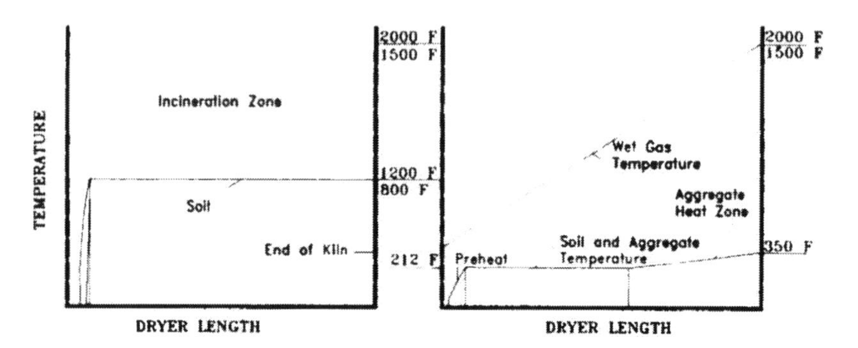

Figure 3-9. Petroleum hydrocarbon-contaminated soil in an evaporative-type unit or batch plant (left) and incineration-type unit (right).

end of the kiln, the raw materials move down an incline to the lower end, where they are heated by coal, oil, or gas (Figure 3-9). The most common raw materials include lime, silica, and alumina; any materials that provide these components can be used provided that they do not contain excessive amounts of other oxides. Contaminated soil may be introduced along with the raw materials or dropped directly into the hot part of the kiln. The raw materials are then heated to high temperatures, ranging up to 2700°F, which causes physical and chemical reactions such as evaporation of free water, evolution of combined water and carbon dioxide from carbonates, and combination of lime with silica, alumina, and iron to form the desired compounds in the clinker. Petroleum-affected soil chemically breaks apart during this process, whereas the inorganic compounds recombine with the raw materials and are incorporated into the clinker. The clinker results in dark, hard nodules the approximate size of a golf ball (3/4-in. in diameter). Portland cement is the result of the rapidly cooled clinker, which is mixed with gypsum and ground to a fine powder.

Portland cement is a hydraulic cement produced by pulverizing clinker consisting of essentially hydraulic calcium silicates, usually containing one or more of the forms of calcium sulfate as an interground addition. Hydraulic cement is the basic binding agent in concrete and masonry construction. Comprising about 95% of the cement produced in the United States, it is a product of high temperature burning of calcarious materials (i.e., limestone, shell, etc.), argillaceous material (i.e., clay), and siliceous material (i.e., sand, shale, etc.) to produce clinker. The four basic ingredients are silica, lime, iron, and aluminum (clay) source). Portland cement thus consists of pulverized clinker, which is blended with water and/or untreated calcium sulfate (gypsum).

Cement manufacturing can be accomplished by one of three processes, wet, dry, and dry with preheating and/or precalcining. These processes differ essentially in the preparation of feed materials prior to calcination. The wet process involves the mixing of finely ground raw materials with water to form a slurry feed. Moisture for this feed ranges from 30 to 40%. Limestone, sand, and clay are interground in wet mills, producing a uniform feed. Aqueous wastes and certain hazardous solid wastes may be added to the raw mill while preparing the slurry.

The dry process involves raw materials that are typically quarried and crushed to an approximate 5-in. diameter. These materials are processed through direct-contact rotary driers to a rotary raw mill, where they are subsequently ground to approximately 200-mesh. Efficiently designed dry-process kilns will consume about 60% of the energy required to produce a ton of cement in a conventional wet process. Raw grinding and preparation of a uniform feed are, however, more difficult than with the wet process. The design of the modern kiln is also more complicated. Thus, process-control challenges are increased.

With the dry process with preheating, this dry powder is introduced to the preheater and passes through a series of heat exchangers prior to entering the kiln. A suspension preheater is situated upstream of the kiln. The preheater consists of a series of cyclones connected by pipes. Gases migrate through the pipes from the kiln and pass upward and countercurrent to the dry raw material flowing downward and around the cyclones. Suspension preheaters essentially transfer the heat from the gas into the raw material feed dust, resulting in approximately 40% calcination of the feed prior to its entering the kiln. Newer preheater systems use a smaller direct-fired furnace located between the air suspension preheater and the kiln. This system can calcinate approximately 90% of the raw material and can reduce the size of the rotary kiln required or increase the production capacity of an existing kiln. In the dry process with precalcining, the precalcining system uses a secondary firing process within the preheater to increase thermal preparation of the feed.

Petroleum-contaminated soil can be processed by being suspended in liquids that are pumped into the hot end of the kiln; they can be injected into the kiln area where gas temperatures range from 1800 to 2150°F, or the solids and sludges can be preprocessed, dried, ground into powder, and conveyed by air into the hot end of the kiln. Regardless of the process used, the ground and blended raw materials are introduced to the rotary kiln, a large, inclined, rotating cylindrical furnace from 10 to 20 ft in diameter and 350 to 760 ft long. Raw materials are introduced into the raised end, traveling down the incline to the other end, which is heated by burning fuel. The retention time in the kiln runs approximately 1 to 4 h. Temperatures at the hot end range from 2500 to 3500°F. The produced clinker cools by air in a clinker cooler. The air from the cooler, along with combustion gases and water vapor, rises and is collected in a dust collection system as it rises to the cool end of the kiln and eventually is emitted out of the stack. The produced clinker is grounded in an open or closed circuit mill. About 3 to 6% gypsum (calcium sulfate) is added to retard the cement setting time. Other additives may be introduced for air-entraining, dispersing, and waterproofing. The final product measures about 10 μm in diameter.

Cement kilns can also accommodate a wide variety of metal-bearing sludges and solid hazardous waste. Solid hazardous waste including iron-bearing slag, residue from acid plants, and other industrial processes have also been successfully used for cement plants with iron-poor mix. Lime sludge derived from water-treatment plants typically contains aluminum and calcium hydroxide, which can be effectively used in the production of cement. Organic waste, including tars, sludges, and plastics, can be effectively destroyed via pyrolysis, with the pyrolyzer gas being used as fuel for the kiln. In addition, depending on the chemical composition of the ash, the ash may become part of the kiln feed. Cement companies in general are not interested in economic recovery as much as they are in raw material replacement.

Undesirable materials to avoid in the cement process fall into two categories, technologically unacceptable and aesthetically unacceptable. Those materials considered technologically unacceptable, which must be evaluated on a case-by-case basis, include excessive amounts of sodium, potassium, chlorides, sulfates, chromium, lead, and other metals. Many soils are commonly high in alkali, which can be deleterious to the process. Excessive amounts of certain constituents can produce adverse operation or production effects, such as unstable kiln operation, poor quality cement, refractory damage, objectionable cement color, or unacceptable safety risks. Aesthetically unacceptable factors include public acceptance and worker safety. Those materials that generate odor problems, such as sewage sludge, sulfides, and mercaptans, and that require special handling may also affect the economics of processing such materials.

Several utilities have evaluated the use of cement kilns for the management of manufactured gas plant contaminated soil and residuals. In Heartland, Kansas, about 50 tons of organic-contaminated soil were handled at a cement kiln as part of a test burn trial. Cement kilns have also been used to handle about 4000 and 1000 tons of petroleum hydrocarbon soil in Wisconsin and Iowa, respectively.

Industrial caliber rotary kilns are located throughout the United States. Some use has been made of petroleum-contaminated soil for daily cover at landfills. Estimated costs vary, depending upon soil and contaminant type; however, estimated costs of $30 to $100 per ton, exclusive of consulting, transportation, and storage, have been reported.

3.2.3 Brick Manufacturing Processes

Contaminated soil has been used as an ingredient in the production of bricks. The contaminated soil essentially is used to replace one or a combination of raw materials, such as shale and/or firing clay. Brick manufacturing incorporates the blending of clay and shale into a plasticized mixture, which is then extruded and molded into brick. When dried, the green-colored brick is fired in a kiln, where temperatures reach approximately 2000°F during a 3-d residence period. With the capability to incorporate contaminated soil with the clay and shale, petroleum-contaminated soil is typically molded into the green brick, dried, and preheated. The brick is fired at 1700 to 2000°F for approximately 12 h in the kiln; the temperature and residence time while the brick is in the kiln destroys the organics, thus incorporating the inorganics in the vitrified brick end product.

Blending of contaminated soil with the clay and shale materials is conducted in stockpiles. Grinders are utilized to reduce these raw materials to a suitable size for brick manufacturing. The plasticity of the blended material is enhanced by the addition of water using a pugmill. A continuous ribbon of clay is extruded by the pugmill, which is subsequently cut into bricks. Produced bricks are typically stacked on rail cars that then travel through a tunnel kiln and pass through three temperature ranges. Initial drying, preheating, and peak heating are conducted at 600°F, 1200 to 1600°F, and 1700 to 2000°F, respectively, the later phase continuing for a period of about 12 h. The travel time for the kiln is about 2.5 d. Upon cooling, the bricks are ready for shipment.

Brick manufacturing facilities that incorporate contaminated soil, notably petroleum-affected soil (including silt, sand, loam, and clay) and fly ash in some cases, are known to exist in California, North Carolina, South Carolina, and Virginia. At a brick manufacturing facility in Ohio, bricks were produced from about 6000 tons of organic-contaminated soil derived from a manufactured gas plant. Other materials include highly plastic clays, along with some shales and sedimentary rocks that can be used for feedstock. Sand, however, will have a tendency to reduce firing shrink and improve moisture absorption from mortar, which is important during brick laying. Estimated costs range from $30 to $45 per ton, excluding consulting, transportation, and storage costs.

3.3 RECLAIM TECHNOLOGIES

Reclaim technologies are in a general sense considered treatment vs. recycling. Such technologies include energy recovery, thermal desorption, solvent extraction, incineration, vapor extraction, and bioremediation (Table 3-2). These technologies have the capability to treat contaminated soil and subsequently dispose of such materials in the areas from which they were excavated or derived, or to reduce the category of risk for landfill disposal. However, neither of them produces a viable, marketable, commercial product.

3.3.1 Energy Recovery

Energy recovery is the use of waste material as fuel in energy-intensive processes (i.e., in the form of steam or process heat). Most commonly utilized in the manufacturing of Portland Cement, a large quantity of combustible waste is burned as fuel in cement kilns each year. More than 25 cement kilns, one-third of all cement plants in the country, currently are permitted and are operational throughout the United States, with an additional ten plants soon

Table 3-2 Summary of Reclaim Technologies

Technology	Contaminants	Media	End product	Limitations
Energy recovery	Organic solvents Organic sludges Petroleum Monomers Wood debris	Flowable soils, sludges, sediments, or particulates	Heating value in boiler, furnance, or cement kiln	Energy content Ash content Impurities Potential toxic by-products Explosive hazard Waste moisture content
Cement kilns	Petroleum hydrocarbons Organic solvents	Soil, sludge, or sediments	Energy value	Energy content Impurities
Indirect heating units	Petroleum hydrocarbons Organic solvents	Soil, sludge, or sediments	Energy value	Energy content Impurities Explosive hazard
Fluidized or rotary kiln	Petroleum hydrocarbons Organic solvents	Soil, sludge, or sediments	Energy value	Energy content Impurities
Solvent extraction	Volatile or semivolatile organics Organic solvents Organic sludges	Soils, sludge, or sediments	Energy value Organic liquid	May produce mixed organic product May require several treatments

Thermal desorption	Petroleum hydrocarbons Organic solvents Organic sludges	Soil, sludge, or sediments	Energy value	May produce mixed organic product Heavy-end hydrocarbons May require several treatments
Soil vapor extraction	Volatile organics	*In situ* vadose zone soils Stockpiled soil	Energy value	Extraction/collection efficiency Minimal effectiveness Low permebility soil May produce mixed organic product
Decanting	Organic solvents Petroleum hydrocarbons Organic sludges	Soil, sludge, or sediments	Organic liquid	May produce mixed organic product Fluid density Impurities
Physical separation	Metals	Soil	Metals	Impurities Moisture content

Modified from United States Environmental Protection Agency, 1994.

to be operational. Growing significantly since the mid-1980s, about 6.8% of all hazardous waste generated in 1995 was used as fuel consumption at cement kilns. According to the U.S. EPA, about 23.6 million t (26 million tons) of hazardous waste fuel, with a heating value greater than 9,000 kJ (8500 BTU), is available, although less than 10% of this fuel is presently used for energy consumption.

Rotary kiln plants can be constructed as permanent installations or as transportable plants. Applicable to solids, sludges, and slurries, rotary kilns are capable of receiving and processing liquids and solids simultaneously. Processing of hazardous wastes with temperatures in the range of 2000°F (1100°C) is readily applicable to kiln technology. Rotary kilns provide a number of functions necessary for incineration by providing for conveyance and mixing of solids, a mechanism for heat exchange, means of ducting the gases for further processing, while serving as a host vessel for chemical reactions.

The process involves the use of raw materials, such as limestone, clay, sand, iron ore, and solid waste of various types. The waste fuel is transported to the cement plant or to a permitted waste fuel processor (or blender) by a broker for the kiln operator. Pretreatment is conducted to prepare the waste fuel for burning. The material is fed into the back or higher end of a rotary kiln either wet or dry (Figure 3-10). Fuel is burned in the lower or front end so that the flow is the same as that of the solids. Toward the front end, the solids are heated, dehydrated, calcined, and combusted and crystallized to form cement clinker.

To be energy-intensive, with maximum gas temperatures in excess of 2200°C (3990°F) toward the front end, much higher energy requirements are needed in comparison to what is typically encountered in hazardous waste incinerators that operate at less than 1480°C (2700°F). Incinerators also maintain relatively shorter gas retention times in comparison to cement kilns, with greater emissions and less fuel combustion.

A wide variety of waste types can be recycled via energy recovery technology, depending on BTU content (heat content), physical characteristics, and chemical composition of the soil. Included is the use of organic-contaminated soil and sludges. Applicable constituents include benzene, toluene, ethylbenzene, total xylenes, mixed aliphatic hydrocarbons, acetone, methly ethyl ketone (MEK), and a variety of chlorinated solvents. The most desirable fuel is liquid, as similar to conventional fuels as is possible, with relatively low chlorine content and moderately high BTU content ranging from 25,666 to 41,900 kJ/kg (11,000 to 18,000 BTU/ib). In addition, the total suspended solids content of liquid should be less than 30% in total volume to prevent plugging of the delivery system. When used, the contaminated soil is blended into raw feed when the organic content is less than 0.1%, or when the feed requires some form of thermal separation or direct feed for preheating. Heat content is typically less than 5000 BTU/lb.

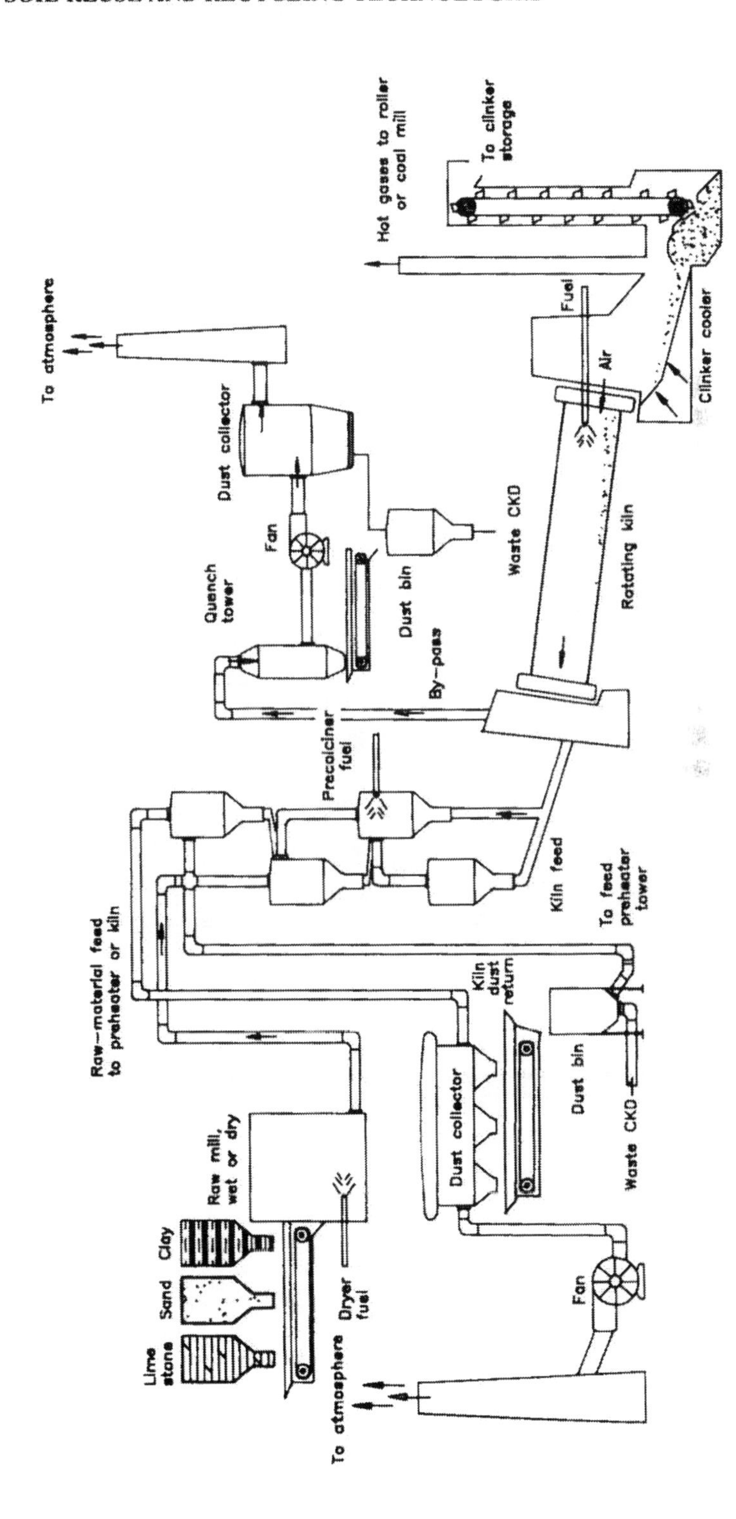

Figure 3-10. Generalized schematic of a cement pyroprocessing plant (note CKD = cement kiln dust).

Table 3-3 Considerations Regarding Tests on Material Processing via Cement Kilns

Test	Purpose
Chemical analysis	Indicates potential endothermic or exothermic reactions.
	Determines quantity of fuel required.
	Determines volume of process gases to be generated.
Specific heat	Indicates amount of heat required to elevate materials' temperatures from ambient levels to required processing temperatures.
Size distribution	Assess potential for dust entrapment in process-gas stream.
	Impacts directly on required residence time at processing temperatures.
Calorific value	Affects fuel quantities.
	Affects process-gas volumes.
Moisture content	Affects fuel quantities.
	Affects process-gas volumes.
	Affects material behavior during early stages of pyroprocessing.
Bulk density	Reflects upon size of kiln and its loading for a given production rate during processing.
Dynamic angle of repose	Affects time required to transport material through kiln.
	Affects residence time.

As with the reuse and recycling technologies, the material to be recycled requires characterization. Characteristics pertinent to rotary kiln technology include chemical composition, specific heat, size consistency or distribution, calorific value, moisture content, bulk density, and dynamic angle of repose. These characteristics of the prospective materials and their respective purposes are summarized in Table 3-3.

3.3.2 Thermal Desorption

Thermal desorption is a process that can be used to physically remove volatile and semi-volatile organic contaminants from soil, in addition to sediments, sludges, and filter cakes, for reuse of the contaminant constituents (Figure 3-11). Included are halogenated and nonhalogenated hydrocarbons, polychlorinated biphenyls (PCBs), pesticides, and dioxins/furans. Due to relatively low temperatures of operation, thermal desorption is best suited for the removal of organics from relatively coarse-grained soil (i.e., sand, gravel, and

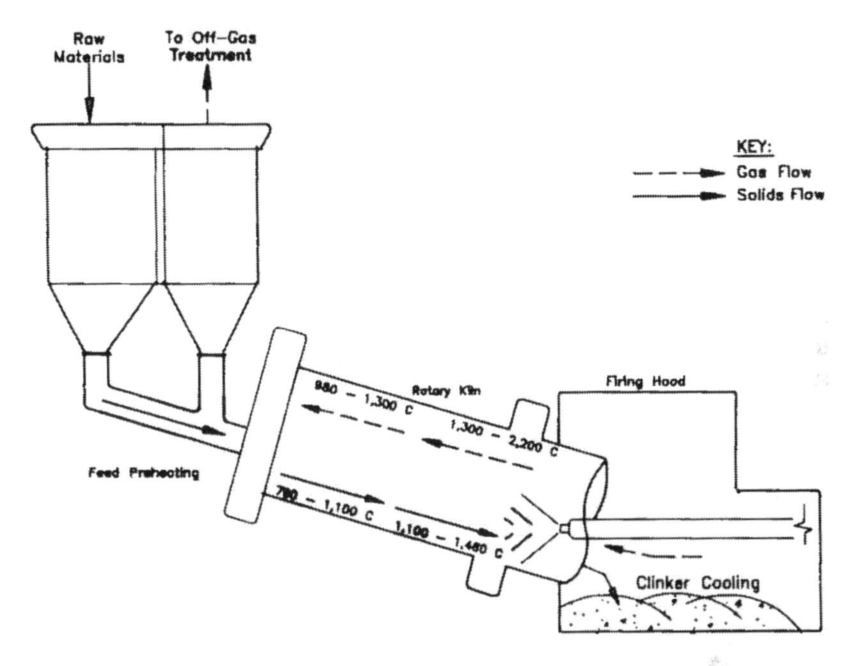

Figure 3-11. Generalized schematic of a cement kiln being used for energy recovery.

rock fragments). Thermal desorption is less suited for relatively fine-grained soil, such as clay, humus, and silt, which are characterized by high-sorption capacity and decreases in the partitioning of organics to the vapor phase.

Thermal desorption systems heat the contaminated material, thus increasing the rate of volatilization of the contaminant, resulting in the organic partitioning to the vapor phase. The mechanisms involve a combination of both volatilization and decomposition. The organic off-gas stream is subsequently collected and processed. Particle sizes acceptable for thermal desorption vary with the particular unit type being used; however, typical size ranges are from 3.8 to 5.1 cm (1.5 to 2 in.) in diameter.

Two types of thermal units exist, indirect heating and fluidized or rotary kiln-type units. Indirect heating units require relatively smaller particles in order to provide sufficient contact surface with the heated wall. Fluidized or rotary-kiln units require a narrow particle size range to control particle residence time in the heat zone. Neither units operate effectively for large fragments, due to heat transfer limitations and the potential for mechanical damage to the equipment.

Thermal desorption equipment is available as modular or trailer-mounted units, which can easily be transported to the site, providing the maximum gross vehicle weight does not exceed 36,300 kg (80,000 lbs.).

Thermal desorption differs from incineration in that it does not oxidize the mineral to its mineral state but, rather, simply removes it from the affected soil. Thermal desorption thus operates at much lower temperatures, ranging from about 95 to 540°C (i.e., 200 to 1000°F). Heating equipment includes rotary kilns, internally heated screw augers, externally heated chambers, and fluidized beds. Some units use two-stage heating, in which the initial low-temperature stage is used for the removal of water while the second stage is used for the removal of organics. Thermal desorption is more suitable for soils or waste with low moisture content, since more heat is required to evaporate water, thus increasing costs and overall effectiveness since the presence of water does not assist in organic partitioning.

An inert carrier gas is used to remove organics from the heated media. Several steps are required for the treatment of the off-gas. These steps involve off-gas conditioning, organic removal, and organic collection. Off-gas conditioning for effective organic collection is achieved by removing organic impurities by the use of cyclone separators and baghouse filters. Organic collection is accomplished by the use of scrubbers and countercurrent washing and condensation. The resultant clean gas is recycled to the heating unit while the discharged portion is cleaned via carbon adsorption.

The advantage associated with thermal desorption is that it is applicable to a variety of organics from complex solid matrices. No combustion products are produced, and the relatively low temperature relative to incineration is less intensive, which also reduces the partitioning of metals to the off-gas if present. The resultant soil following treatment can be used for recycling via asphalt incorporation or can be injected into a cement kiln or furnace for energy recovery.

Thermal desorption depends on the ability to control and maintain the heating of the affected soil; treated material almost always retains traces of organic contaminants. In addition, the stripped organics must be distilled or purified prior to being reused as a solvent. Dust abatement may be required during processing and when the material is transferred out of the heating unit. In regard to cost, thermal desorption is capital-intensive, and pricing for treatment is typically on a unit cost basis, regardless of how many times the material may have to be processed prior to cleanup goals being achieved; the product may still be considered a waste that requires special handling.

3.3.3 Solvent Extraction

Solvent extraction involves the use of an organic solvent in contact with the contaminated material to recover organic contaminants from soil, sludges, and sediments (and liquids) for reuse of the contaminant (Figure 3-12). Typical solvents include liquefied gas (propane or butane), triethylamine, or proprietary organic fluids. The extraction solvent is thoroughly mixed with the contaminated matrix, allowing transfer of contaminants to the solvents. The clean

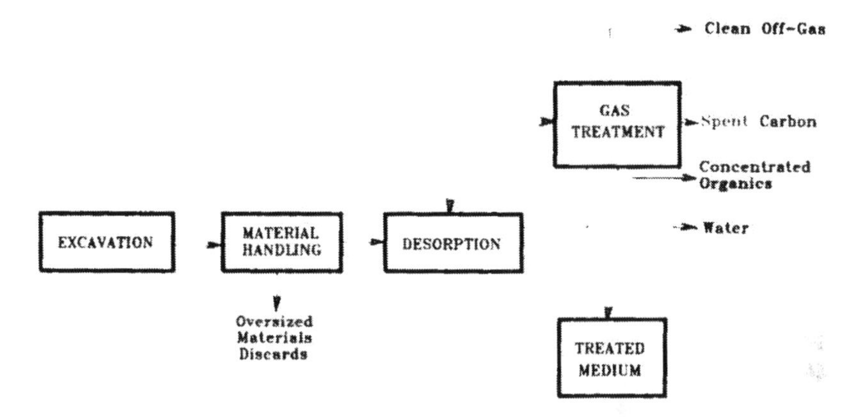

Figure 3-12. Schematic of the thermal desorption process.

matrix and solvent are then separated via gravity, decanting, or centrifuging. Distillation regenerates the solvent for reuse, or it is processed and used for energy recovery.

Wastes applicable for solvent extraction include volatile organic compounds, PCBs, halogenated solvents (i.e., pentachlorophenol), and petroleum. Extraction is more effective with lower molecular weight hydrophobic compounds.

In solvent extraction, the extraction solvent should be characterized by high solubility in the contaminant and low solubility in the waste matrix. Some systems require the addition of water if the soil is dry and nonflowing. Other systems require the addition of extraction fluid to make the waste flow. The extraction solvent is subsequently purified by distillation. In systems using pressurized solvents, such as liquefied gas or supercritical carbon dioxide, vaporization occurs through pressure release, which causes the solvent to boil. Distillation tanks or towers are used with higher-boiling solvents to separate the extraction solvent from the organic contaminants.

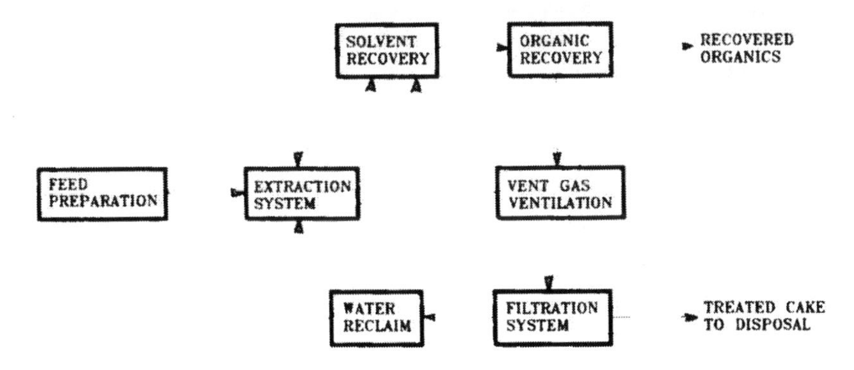

Figure 3-13. Schematic of the solvent extraction process.

The triethylamine system extracts both water and organics by heating the contaminant/water/solvent mixture to 55°C (130°F), resulting in separate water and organic waste forms. These phases are then separated by decanting, whereas, the contaminant and solvent are separated by distillation.

As with other reuse technologies, solvent extraction recovers organic contaminants from an inorganic matrix, thus reducing the volume of contaminated soil requiring special handling and preparing the contaminant for recycling. Another advantage is that this technology can be used to treat wastes with high concentrations of organic contaminants. However, most extraction solvents are volatile and flammable, and require precautions to reduce fire and explosion risks. In addition, if the affected soil contains elevated metals, organically bound metals can transfer to the extraction solvent with the organics, thus restricting reuse options.

Lower grade uses, notably of mixtures of nonchlorinated organics, may be suitable for energy recovery or asphalt incorporation. For higher grade uses, notably when chlorinated organics are present, further processing may be required. Detergents and emulsifiers in the waste can reduce overall extraction performance, promote foam formation complicating separation of the matrix and extraction solvent, and in the case of water soluble detergents, dissolve and retain organic contaminants in the matrix.

3.3.4 Soil Vapor Extraction

Soil vapor (or vacuum) extraction (SVE) is typically used as an *in situ* remedial technology for the removal of volatile organics from the vadose (unsaturated) zone. Vapor extraction is also used to enhance light nonaqueous phase liquid (LNAPL) recovery when minimal LNAPL occurs or when such occurrence is in low permeability soil, a process referred to as vacuum-enhanced extraction.

SVE systems utilize blowers attached to vapor extraction wells. Extraction wells are constructed in a similar manner as groundwater monitoring wells, although they seldom penetrate the water table and are screened across hydrocarbon-affected zones. Air injection wells are sometimes drilled to induce air flow through the soil matrix, which strips the volatile organics from the soil and induces migration of the volatile organics to the extraction wells. Partitioning of volatile material from either solid, dissolved, or NAPL phases occurs with the introduction of blower-generated air. The resulting air emissions are then controlled via adsorption onto activated carbon, thermal destruction via incineration, or catalytic oxidation, or by condensation via refrigeration.

Although no bulk excavation is required, separation of various organic species may be required when a mixture of organics is present (i.e., fuel hydrocarbons and chlorinated solvents). When relatively high concentrations are present, a sufficient volume of liquid is retrieved to serve as a fuel for operation of the SVE system or to be cleaned and reused in the case of LNAPLs.

3.3.5 Decanting

Decanting is a process for physically separating two immiscible liquid phases, thus allowing for purification and reclamation or reuse of one or more of the phases. Typically used to remove small quantities of water dispersed in oil or vice versa, the process relies on gravity to separate the dense from the light liquid phases.

A decanter is essentially a tank with a large surface area to volume ratio. The continuous phase resides in the tank, whereas the dispersed droplets phase combine and rise or sink, depending on density, to form a separate phase that can be decanted. Decanting is most efficient when the surface area available for development of a second phase is large in comparison to the volume of the fluid. Parallel separators or plates are commonly used in order to provide a large surface area to volume ratio.

When the continuous and dispersed phases are difficult to separate or when a high proportion of solids exists, accessory units, such as air-flotation, centrifuges, coalescers, or hydrocyclones, may be utilized. Air-flotation units are used mainly for low-density hydrophobic materials such as oil or when the dispersed phase concentration is low. Air-flotation units can improve phase separation by introducing bubbles into the continuous phase. The bubbles provide a large surface area, which serves to collect the dispersed phase objects. The formation of bubbles is accomplished simply by saturating the water with air at elevated pressure, then releasing the pressure. Bubbles can also be introduced via gas sparging or electrolysis. Centrifuging units are used to separate light oil, water, and solids by mechanically spinning. Hydrocyclones, on the other hand, accomplish separation via radial flow. Coalescer units provide sufficient surface area that is used to enhance contact and agglomeration of disposed phase droplets. The surface can consist of a packed bed, a fiber mesh, or a membrane.

Applicability to contaminated soil is limited, but decanting has been used to remove oil from oil sludges, although stable emulsions and suspensions must usually be disrupted in order to allow sufficient physical separation. Decanting is limited to the separation of immiscible liquids such as oil and water and is applicable to miscible liquids such as mixtures of organics.

3.3.6 Physical Separation

Physical separation techniques are available to concentrate, recover solids suspended in water or mixed with other solids (i.e., reclamation of metals or mercury from soil or lead from firing-range soil), or determine the specific size fractions within a soil matrix in which certain contaminants may reside. Most applicable to metal-contaminated soil, some of the more conventional particle separation techniques include sieve analysis (or particle size distribution), settlement velocity, gravity separation, magnetic separation, and flotation. Sieve analysis relies on varying particle size with screens sized with various diameter openings, allowing the passage of particles of varying effective sizes by use of screens, sieves, or trammels. Separation via settlement velocity relies on differences in particle density, size, or shape. Particles are separated with the use of a clarifier, elutriator, or hydrocyclone. Separation via gravity relies on density differences and is performed using a shaking table, spiral concentrator, or jig. Magnetic separation relies on magnetic susceptibility, using electromagnets or magnetic filters. Particle separation by flotation relies on the particle attraction to bubbles due to their surface properties. Flotation separation is accomplished through the use of air flotation columns or cells.

BIBLIOGRAPHY

Ahmed, J., 1993, Use of Waste Materials in Highway Construction, Noyes Data Corporation, Park Ridge, NJ.

Ali, M., Larsen, T. J., Shen, L. D., and Chang, W. F., 1992, Cement stabilized incinerator ash for use in masonry bricks — cement industry solutions to waste management, in Proceedings of the First International Conference, Calgary, Alberta, Canada, October, 1992, pp. 325–331.

The Asphalt Institute, 1982, Principles of Construction of Hot-Mix Asphalt Pavements, The Asphalt Institute, College Park, MD, Manual Series No. 22.

The Asphalt Institute, 1983, Asphalt Cold-Mix Recycling, The Asphalt Institute, College Park, MD, Manual Series No. 21 (MS-21), 68 pp.

Bouse, E. F., Jr. and Kamas, J. W., 1988, Waste as kiln fuel, Part II, Rock Products, Vol. 91, pp. 59–64.

Bouse, E. F., Jr. and Kamas, J. W., 1988, Update on waste as kiln fuel, Rock Products, Vol. 91, pp. 43–47.

Chadbourne, J. F., 1988, Cement kilns, in *Standard Handbook of Hazardous Waste Treatment and Disposal* (Edited by H. M. Freeman), McGraw-Hill Book Company, New York, pp. 857–875.

Collins, R. E. and Luckevich, L., 1992, Portland cement in resource recovery and waste treatment — cement industry solutions to waste management, in Proceedings of the First International Conference, Calgary, Alberta, Canada, October, 1992, pp. 325–331.

Czarnecki, R. C., 1989, Hot mix asphalt technology and the cleaning of contaminated soil, in *Petroleum Contaminated Soils*, Vol. 2 (Edited by E. J. Calabrese and P. T. Kostecki), CRC/Lewis, Boca Raton, FL, pp. 267–277.

DuGuay, T., 1993, Reclaim metals to clean up soils, *Soils*, March, 1993, pp. 28–33.

Dulam, C. J., Hoag, G. E., Dahmani, A., and Nadim, F., 1995, A feasibility study to use coal tar contaminated soil in asphalt cement mixture production, in *50th Purdue Industrial Waste Conference Proceedings*, Ann Arbor Press, Chelsea, MI, pp. 357–363.

Eklund, K., 1989, Incorporation of contaminated soils, *Petroleum-Contaminated Soils*, Vol. 1 (Edited by P. T. Kostecki and E. J. Calabrese), CRC/Lewis, Boca Raton, FL, pp. 191–210.

Elliott, E. J. And Brashears, D. F., 1991, A critical assessment of asphalt batching as a viable remedial option for hydrocarbon-contaminated soils, in *Hydrocarbon Contaminated Soils*, Vol. 1 (Edited by E. J. Calabrese and P. T. Kostecki), CRC/Lewis, Boca Raton, FL.

Gossman, D., 1992, The reuse of petroleum and petrochemical waste in cement kilns, Environmental Program, Vol. II, No. 1, pp. 1–6.

Hazardous Waste Consultant, 1994, Alternatives for remediation lead/hydrocarbon-contaminated soils — asphalt incorporation shows promise, 1994, The Hazardous Waste Consultant, July/August, 1994, pp. 1.15–1.20.

I.T. Corporation, 1991, Assessment of Selected Technologies for Remediation of Manufactured Gas Plant Sites, Electric Power Research Institute Research Report Project 3072-1, Report No. GS-7554, October 1991.

Knowlton, R. C., 1992, Remediation of petroleum contaminated soils utilizing asphalt emulsion stabilization technology at the site of generation, in *Hydrocarbon Contaminated Soils*, Vol. II (Edited by P. T. Kostecki, E. J. Calabrese, and M. Bonazomtas), CRC/Lewis, Boca Raton, FL, pp. 511–528.

Kostecki, P. T., Calabrese, E. J., and Fleischer, E. J., 1989, Asphalt batching of petroleum contaminated soils into bituminous concrete, in *Petroleum Contaminated Soils*, Vol. 1 (Edited by P. T. Kostecki and E. J. Calabrese), CRC/Lewis, Boca Raton, FL, pp. 175–186.

Meegoda, N. J., 1992, Use of petroleum contaminated soils in asphalt concrete, in *Hydrocarbon Contaminated Soils*, Vol. II (Edited by P. T. Kostecki, E. J. Calabrese, and M. Bonazomtas), CRC/Lewis, Boca Raton, FL, pp. 529–548.

Meegoda, N. J., 1994, Contaminated soils in highway construction, in *Process Engineering for Pollution Control and Waste Minimization* (Edited by P. L. Wise and D. J. Trantolo), Marcel Dekker, New York, pp. 663–684.

Richardson, K. E., et al., 1993, Technology selection for remediation of lead and hydrocarbon contaminated soil, in Proceedings of the Ninth Annual Environmental Management and Technology Conference West/Fall HazMat 93 West.

Schaefer, C. F. and Albert, A. A., 1988, Rotary kilns, in *Standard Handbook of Hazardous Waste Treatment and Disposal* (Edited by H. M. Freeman), McGraw-Hill, New York, pp. 819–830.

Sciarrotta, T. C., 1989, Recycling of petroleum contaminated soils in cold mix asphalt paving materials, in *Hydrocarbon Contaminated Soils and Groundwater* (Edited by P. T. Kostecki and E. J. Calabrese), Vol. I, pp. 239–251.

Swearingen, D. L., Jackson, N. C., and Anderson, K. W., 1992, Use of Recycled Materials in Highway Construction, Washington State Department of Transportation, Report No. WA-RD 252.1.

Testa, S. M. and Patton, D. L., 1993, Resource recovery of lead-affected soil via asphaltic metals stabilization (AMS), in Proceedings of Superfund XIV, Washington, D.C., pp. 1219–1231.

Testa, S. M. and Patton, D. L., 1992, Paving market shows promise, *Soils*, pp. 9–11.

Testa, S. M. and Patton, D. L., 1994, Soil remediation with environmentally processed asphalt, in *Process Engineering for Pollution Control and Waste Minimization* (Edited by P. L. Wise and D. J. Trantolo), Marcel Dekker, New York, pp. 297–309.

United States Environmental Protection Agency, 1992, Potential Reuse of Petroleum-Contaminated Soil, A Directory of Permitted Recycling Facilities, Report No. EPA/600/R-92/096, 38 pp.

United States Environmental Protection Agency, 1994a, Handbook — Recycling and Reuse of Material Found on Superfund Sites, Report No. EPA/625/R-94/004, 84 pp.

United States Environmental Protection Agency, 1994b, Innovative Site Remediation Technology — Solidification/stabilization, Report No. EPA 542-B-94-001, June, 1994, Vol. 4.

Yan Tsoung-Yuan, 1986, Manufacture of road-paving asphalt using coal tar, Industrial Engineering and Chemical Production, Reservoir Development, Vol. 25, pp. 637–640.

4 FIELD CONSIDERATIONS

4.1 INTRODUCTION

For years, excavations associated with remedial activities have proceeded according to the logic "the more the merrier," as the primary justification for removing excessive amounts of so-called contaminated soil. The area of concern is typically delineated with soil borings or test pits, and soil samples are retrieved for subsequent chemical testing. The soils are then excavated and hauled, processed, or treated. Minimal, if any, consideration is given to geologic factors or chemical distribution of the contaminants in the subsurface. Even less thought is given to the quantity of nonaffected soil that may be handled as contaminated, thus being guilty merely by association. This often results in more nonaffected soil being handled as contaminated than the actual volume of contaminated soil that is present. This method, commonly referred to as "dig and haul," all too frequently results in run-away project costs as excessive volumes of soil are excavated.

Presented in this chapter is discussion of certain field considerations and techniques pertinent to the reuse and recycling of contaminated soil; their implementation may prove effective in minimizing the volume of soil that must be excavated the actual volume of soil that requires special handling, treatment, and/or reuse and recycling. Included is discussion of sampling strategies, particle size distribution and its relationship to selective excavation, and volatile organic compound mitigation.

4.2 SAMPLING STRATEGIES

Retrieval of representative soil and aggregate samples from stockpiles and transportation units is a common practice during the assessment- and segregation-related functions. Unbiased sampling from such units can be difficult due to segregation of differing-sized fractions, which often occurs in stockpiles, where coarser-grained particles end up along the outside perimeter of the base of stockpiles or along the inside perimeter of containment structures. The procedure for retrieval of samples from stockpiles will depend upon the

Figure 4-1. Retrieval of representative soil samples at various intervals from numerous stockpiles.

availability of power equipment (Figure 4-1). When not available, representative samples should be retrieved at a minimum of three increments, the top third portion of the stockpile, the mid-point, and the bottom third of the stockpile. With fine-grained soil or aggregate, the sample should be obtained from beneath the outer layer, which may have become segregated. In addition, sampling tubes can be employed for insertion into the stockpile at random locations. Such tubes are typically 1.25 in. (30 mm) in diameter and 6 ft (2 m) in length. A minimum of five increments of material should be retrieved in this manner for compositing into one sample for subsequent chemical testing.

With the use of power or heavy equipment, an effort should be made to prepare a separate, small sampling pile composed of material derived from various levels and locations from within the main stockpile, after which several increments can be composited for the field sample (Figure 4-2). Should it be necessary to document the degree of variability within the main stockpile, then separate samples representative of separate areas of the stockpile should be retrieved.

When obtaining samples of soil or aggregate from transportation units such as rail cars, barges, or trucks, using equipment capable of exposing materials at various intervals and locations is preferred. When these are not available, a common procedure involves the excavation of three or more trenches across the unit at points that will provide a reasonable visual estimate of the materials characteristics. The trench bottom should be level and at least 1 ft in width and depth below the surface (Figure 4-3). Samples should be obtained from a minimum of three increments from approximately equally

Figure 4-2. Retrieval of representative soil samples at various intervals from one large stockpile.

spaced points along each trench and at a minimum depth of 1 ft below the bottom of the trench.

4.3 PARTICLE SIZE DISTRIBUTION

Particle size distribution (PSD) is employed to minimize the quantity of excavated soil that requires special handling and processing by separating in

Figure 4-3. Retrieval of representative soil samples from trenched area.

the field those size fractions that are contaminated vs. those noncontaminated. This determination is made by subjecting representative samples of the site soil matrix to gradation or sieve size analysis in accordance with ASTM C 136. PSD is based on the premise that some types of contaminants mobilize and tend to migrate and concentrate within a given size fraction. Each sieve size, from over 2-in. through #200 minus, is examined, sampled, and analyzed from an engineering and chemical perspective to determine the affected particle size ranges. Once identified, affected particle size ranges can then be mechanically separated from nonaffected sizes — which reduces the volume of soil to be processed and thus reduces overall project costs. Relatively impervious material, such as rock, concrete, gravel, and rubble, are mechanically separated during this process to further reduce the quantity of soil to excavate and to provide additional cost savings.

For example, a sandy clay matrix, presented in Table 4-1 and illustrated in Figure 4-4, shows total lead concentrations generally increasing with decreasing grain size, with a significant increase equal to or less than the 4-mm size fractions.

Each contaminant type or chemical group reacts differently under different soil conditions, and each migrates and concentrates in the subsurface in a distinctive manner, depending on grain size, permeability, and porosity. Structural elements related to the depositional environment, such as stratification, bedding, and lateral–vertical facies changes also play an important role in the distribution of contaminants in the subsurface. Usually metals are

Table 4-1 Particle Size Distribution for Lead

Size fraction	TTLC (mg/kg)	STLC (mg/l)	TCLP (mg/u)
1½	30.9	NA	NA
1	186.0	1.14	ND (<0.20)
¾	120	2.18	ND (<0.20)
½	27.3	NA	NA
⅜	119	0.74	ND (<0.20)
No. 4	276	1.49	NA
No. 8	246	1.88	NA
No. 16	237	2.84	NA
No. 30	277	4.00	NA
No. 50	359	3.09	NA
No. 100	413	3.81	NA
No. 200	235	3.15	NA
200 Minus	329	3.77	NA

Note: NA = Not analyzed; ND = not detected at or exceeding the analytical detection limit as shown in parentheses.

Figure 4-4. Test pit showing limited vertical extent of metal slag-contaminated soil.

more constrained relative to hydrocarbons and do not tend to migrate laterally or vertically downward to any significant distance (Figure 4-5).

4.4 SELECTIVE EXCAVATION

Selective excavation is a procedure to reduce the overall quantity of affected soil to be excavated and handled to its lowest quantifiable amount in order to reduce project costs. Selective excavation uses controlled excavation techniques and sampling to segregate soil within the impacted area into affected and nonaffected classifications. Selective excavation is performed on a horizontal plane. The site is laid out in a grid pattern, as illustrated in Figure 4-6, with alphabetic designations noted left to right and numeric designations top to bottom. Depending on site conditions, contaminants of concern and anticipated plume delineation, spacing will vary from 5 to 50 ft or more apart. All related analytical data from samples are referenced by their location relative to the intersecting grid lines.

Site preparation is site specific and may include demolition, construction of access routes, preparation of water access points, installation of materials handling conveyors, installation of protective fencing, and basically whatever is required to meet regulatory requirements and expedite the project. The final preparatory step is to sample all intersecting grid points at the surface or to 1-ft depth. The samples are analyzed with chemical field screening or a mobile lab and submitted for laboratory confirmation of field results, allowing the accurate documentation and correlation of field to laboratory data. Soil is removed in predetermined lifts and stockpiles according to its "affected" or

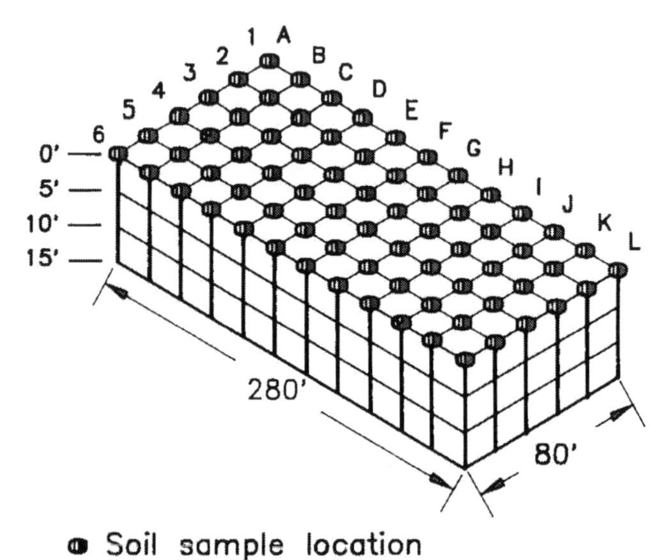

Figure 4-5. Schematic showing approximate soil sample locations during conduct of selective excavation.

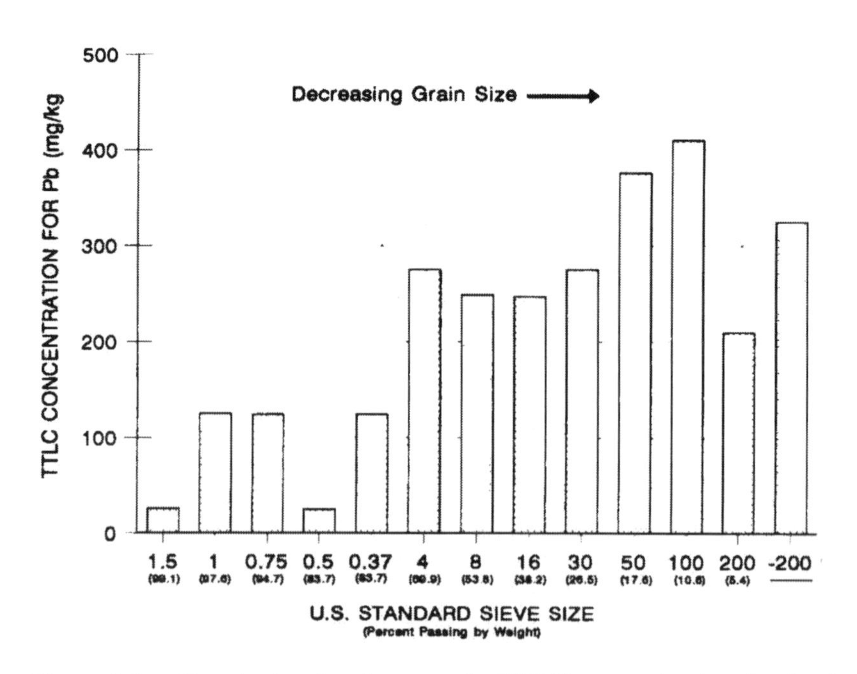

Figure 4-6. Graph showing increase in TTLC concentration for lead with decreasing grain size for a sandy clay at a steel manufacturing plant.

"nonaffected" classification. This classification is determined through communication with the local lead regulatory agency (as, so many ppm gasoline-affected soil or so many ppm diesel-affected soil). To expedite the process, the area of excavation can be divided into quadrants, where excavation occurs in one quadrant while sampling and testing takes place in another. Sampling and field analysis commences while excavation recommences in a quadrant previously sampled and analyzed.

During the course of excavation, all intersecting grid points are sampled and analyzed. Wooden grade stakes, color coded red for affected and green for nonaffected, are driven into the grid points after sampling. This process is repeated for every lift removed. The delineation of the plume typically becomes visually apparent at this point. Little if any time is lost excavating materials in a quadrant-to-quadrant manner. Affected soil, identified by the red stakes, is placed outside the excavation and transferred by loader or conveyor to the affected stockpile. Nonaffected soil is classified, excavated, and stockpiled in the same manner. Affected soil can be mechanically separated by on-site vibratory screening units to remove rock, debris, and other impervious, nonaffected oversize materials. The oversize materials may be used as backfill or processed through a rock crushing unit to produce usable aggregate.

Highly reliable field infrared and soil vapor screenings are performed on samples that are obtained from both stockpiles. An on-site mobile laboratory may be used where the type of contaminant permits. When necessary, laboratory confirmation samples are sent for rush turnaround to get results as soon as possible. Stockpiles are constructed to positively delineate each day's production of soil to avoid misclassification.

Initial implementation of particle size distribution methods can help to further pinpoint the specific soil type and grain size affected, which can then be mechanically separated during excavation activities. Use of selective excavation combined with particle size distribution methods can significantly reduce the volume of affected soil that must be handled, resulting in cost savings. Nonaffected soil, which would typically be handled as contaminated soil, can instead be used as backfill, which further reduces project costs by reducing the volume of imported clean soil required to restore the site.

4.5 VOLATILE ORGANIC COMPOUND MITIGATION

With the use of water-based emulsion with CMA processes, no significant vapors are associated with the emulsion being used for incorporation of contaminated soil into CMA. However, asphalt pavement, notably hot-mix asphalt, is known to emit tons of reactive organic gases per day (i.e., in the South Coast Basin of southern California, over 5 tons of reactive organic gases are released daily), which contribute to smog-forming reactions. These organic gases include a wide spectrum of contaminants that can be broadly classified as polycyclic aromatic hydrocarbons (PAHs) and volatile organic compounds (VOCs). Some PAHs, such as anthracene, phenanthrene, pyrenes, and benzopyrenes, are well-known carcinogens. Emissions associated with asphalt are typically highest during paving, although long-term emissions are also known to occur since the asphalt surface is an ideal solar heat absorber, thus gradually releasing volatile organics with time. In addition, hydrocarbon-impacted soil being excavated or stockpiled may emit elevated volatile or semivolatile organic vapors, which may pose a health and safety hazard or may be aesthetically detrimental to the work being performed.

Volatile organic compound mitigation (VOCM) is a field procedure used to limit and control the airborne emissions of such compounds during all phases of excavation, classification, stockpiling, and processing. Potential hydrocarbon emissions are routinely monitored continuously or at a minimum of every 20 min during conduct of field-related activities. When elevated emissions from the soil being excavated or the soil stockpile exist, then application of water, vapor suppressants, penetrants, or surfactants, or a combination thereof can be used with no significant adverse impact on the quality of the end product.

During conduct of field activities, air monitoring is performed. With the use of hand-held sprayers (i.e., Hudson-type sprayers), the vapor reduction compound is sprayed on the stockpile, excavation, or trench wall, or on newly exposed areas during excavation. VOC readings are routinely taken at 5-min intervals until vapor elevations are stabilized. When the contaminated soil includes ignitibles (i.e., gasoline), airborne concentration of the ignitible compounds should be limited to less than 25% of the lower explosive limit (LEL) during processing.

BIBLIOGRAPHY

Testa, S. M. and Patton, D., 1993, Don't dig clean soils — selective excavation can cut project costs in half, *Soils*, December, 1993, pp. 31–33.

5 LABORATORY CONSIDERATIONS

5.1 INTRODUCTION

The overall effectiveness of reuse and recycling efforts as it pertains to the incorporation of contaminated soil into asphaltic or concrete end products is measured in terms of leachability. From an environmental perspective, leachability is the most important parameter to be considered. Leachability testing measures the potential of an end product to release contaminants or constituents of concern to the environment. As with all these types of tests, the soil, waste, or end product is exposed to a leachant, and the concentration of contaminant potentially released in the leachate or extract is determined. The resultant concentration is then compared to some established regulatory standard.

Presented in this chapter is discussion of hazardous waste types and characterization considerations in order to meet certain acceptance criteria. Leachability and the various types of leachability tests available and more commonly used pertinent to reuse and recycling of contaminated soil into asphaltic or cementitious end products are also discussed.

5.2 HAZARDOUS WASTE TYPES

Waste can generally be categorized into 20 waste types based on their respective physical and/or chemical form, and the hazardous constituents contained within the waste. Examples of such waste types are presented in Table 5-1. These generalized waste types are essentially equivalent to the SARA categories, with the exception of three additional definitions.

Halogenated solvents. Any liquid waste that contains an organic constituent in the F001 to F005 definitions, has greater than 90% organic content, as well as greater than 0.1% halogen content. Halogen content refers to organic halogen content as opposed to

Table 5-1 Examples of Generalized Waste Types

Waste type	Examples
Halogenated solvents	Methylene chloride, chloroform, trichloroethylene, carbon tetrachloride, perchloroethylene
Nonhalogenated solvents	Acetone, methanol, isopropyl alcohol, butanol, ethyl ether, pyridine, benzene, toluene, xylene, hexane, methyl ethyl ketone, ethyl acetate
Halogenated organic liquids	Waste from 1,1,1-trichloroethane production
Nonhalogenated organic liquids	Ignitable wastes, distillate bottoms from phenol, acetone, or aniline production
Organic liquids, NEC	Waste oils, phenol, cresol, formaldehyde
PCB/dioxin	Arochlors, askarels, transformer oils
Pesticides/herbicides	Chlordecone, dicofol, endosulfan, methyl bromide, diazinon, parathion, malathion, diquat, diuron, paraquat
Mixed organics with liquids	Incinerable organic liquids with inorganics (high organic content)
Inorganic liquids with organics	Aqueous or nonaqueous with organics (low organic content)
Inorganic liquids with metals	Wastewaters with cyanide or chrome
Inorganic liquids, NEC	Sulfuric acid, hydrochloric acid, nitric acid, caustic scrubbers, sodium hydroxide, caustic soda solutions, bleach
Halogenated organic solids/sludge	Still bottoms, heavy ends from methylene chloride, trich, perc
Nonhalogenated organic solids/sludge	Still bottoms, heavy ends from acetone, benzene, pyridine
Organic solids/sludge	Emulsion solids, tank bottoms, crude NEC, oil sludge, grease
Dyes/paints/resins	Paints, paint thinners, resins, tars, acrylics, phenolics, lacquers, varnish, coatings
Mixed organic/inorganic solids/sludge	Incinerable solids/sludges with inorganics (high organic content)
Inorganic solids/sludge with metals	Metal salts, filings, scale, oxides, spent catalysts
Inorganic solids/sludge NEC	Carbon, clay, filter cartridges
Contaminated soils	Soil, sand, clay, bricks, gravel, rubble, glass, grass, trees
Other wastes	Explosives, low radioactive waste, materials contained in lab packs, containerized gas

Modified from California Environmental Protection Agency, 1992, Commercial Hazardous Waste Facilities for Recycling/Recovery Treatment, Disposal (Draft), Document No. 902, 97 pp.

inorganic halogen salts, such as sodium chloride. To be included in this category are solvents whose halogen content has not been determined.

Nonhalogenated solvents. Any liquid waste that contains an organic constituent in the F001 to F005 definitions, has greater than 90% organic content, and less than 0.1% halogen content. Includes oxygenated and hydrocarbon solvents.

Organic liquids, NEC. Any liquid waste for which nothing is known except that its organic content is thought to be greater than 90%.

Halogenated organic liquids. Any liquid waste that does not contain a constituent listed in the F001 to F005 definition, has greater than 90% organic content, and has greater than 0.1% halogen content.

Nonhalogenated organic liquids. Any liquid waste that does not contain constituents in the F001 to F005 definition, has greater than 90% organic content, and contains less than 0.1% halogen content.

Inorganic liquids with organic. Any liquid waste that has an organic concentration up to 1.0%, but no metals exceeding 1 ppm.

Mixed organic/inorganic liquids. Any liquid waste that has an organic content between 1 and 90%, regardless of the halogen or solvent concentration.

Inorganic liquids with metals. Any inorganic liquid waste that contains RCRA-regulated metals in excess of 1 ppm and is not thought to contain any organic beyond trace amounts.

Inorganic liquids, NEC. Any inorganic liquid with either unknown constituents, reactive constituents such as cyanide or sulfide, or both metals in excess of 1 ppm and organic up to 1%.

Halogenated organic sludge/solids. Any waste that has greater than 3% total suspended solids, is greater than 90% organic compound, and has greater than 0.1% halogen content.

Nonhalogenated organic sludge/solids. Any waste that has greater than 3% total suspended solids, is greater than 90% organic compound, and has less than 0.1% halogen content.

Organic sludge/solids, NEC. Any waste for which nothing is known except that it is believed to have greater than 3% total suspended solids and to have 90% or greater organic content.

Mixed organic/inorganic sludge/solids. Any waste with greater than 3% total suspended solids and with an organic content between 1 and 90%.

Inorganic sludge/solids with metals. Any waste with at least 3% total suspended solids, at least 10 ppm or RCRA-regulated metal, and that is not thought to contain organics beyond trace amounts.

Inorganic sludge/solids, NEC. Any waste with total suspended solids of 3% or greater and other characteristics are unknown, that is reactive due to cyanide or sulfide, or that contains both metals in excess of 10 ppm and organic up to 1%.

Contaminated soils. Any waste that is primarily soil contaminated with hazardous waste but does not include spent filter media.

Other wastes. Any waste that is explosive, highly reactive, contaminated with low level radioactivity, or contained in lab packs. Also included are state hazardous wastes not covered by RCRA, containerized gases, or any waste where not enough characteristics are known to place it in any of the NEC categories. This category does not include infectious wastes.

The last three generalized waste types below are not SARA definitions but are included here because of their importance to specific industry groups.

PCBs/dioxins. Any waste contaminated with more than 50 ppm of polychlorinated biphenyls (PCBs) or dioxins.

Paints/resins. Solvent- or water-based paints with metal pigments, paint waste, latex waste, adhesives, resins, dyes, printing inks, etc.

Pesticides/herbicides. Pesticide or herbicide off-spec, surplus, canceled formulations or waste; does not include pesticide or herbicide rinse water.

5.3 HAZARDOUS WASTE CHARACTERIZATION

The determination as to whether a waste or solid waste is a hazardous waste is addressed under 40 CFR Parts 261 and 261.4 and Subpart C of Part 261. A hazardous waste is defined as any waste or combination of wastes that poses a substantial present or potential hazard to human or living organisms because it (1) is nondegradable or persistent in nature, (2) can be biologically magnified, (3) can be lethal, or (4) can cause or tend to cause detrimental cumulative effects. A material, substance, or liquid is classified as hazardous if it is listed as a hazardous waste or meets certain specific characteristics. Waste is thus referred to as either a listed or characteristic waste. Listed wastes may exhibit one or more of the specific characteristics and are tabulated in the federal regulations under 40 CFR Part 261, Subpart D. Characteristic hazardous or toxic wastes are determined by evaluating their ignitability, corrosivity, reactivity, and toxicity. These characteristics and their respective properties are summarized in Table 5-2.

Another important waste type is radioactive waste. Radioactive waste and affected soil are ubiquitous throughout the mining and nuclear industries. Radioactive waste can be categorized as either high-level [or transuranic (TRU)] or low-level waste. High-level radioactive waste is restricted to spent fuel rods and the liquid and solid material that can be directly derived from them. TRU waste is material containing appreciable amounts of the heavy actinide elements (chiefly Np, Pu, and Am), which is derived from the production and handling of plutonium for military purposes. Low-level radioactive

Table 5-2 Characteristics of Hazardous Waste

Characteristic	EPA hazardous waste no.	Properties
Hazardous waste ignitability	D001	A liquid, other than an aqueous solution, containing less than 24% alcohol by volume and with a flashpoint less than 60°C (140°F); not a liquid but capable under ambient conditions of causing fire through friction, absorption of moisture, or spontaneous chemical changes and when ignited burns vigorously and persistently; it is an ignitable compressed gas as defined under 49 CFR 173.300; it is an oxidizer as defined under 49 CFR 173.151.
Corrosivity	D002	Aqueous with a pH less than or equal to 2, or greater than or equal to 12.5; liquid that corrodes steel (SAE 1020) at a rate greater than 6.35 mm (0.250 in.) per year at a test temperature of 55°C (130°F).
Reactivity	D003	Normally unstable and readily undergoes violent change without detonating; reacts violently with water; forms potentially explosive mixtures with water; generates toxic gases, vapors, or fumes in sufficient quantities to present danger when mixed with water; a cyanide or sulfide-bearing waste which, when exposed to pH conditions between 2 and 12.5, can generate toxic gases, vapors, or fumes in sufficient quantities to present a danger; readily capable of detonation or explosive decomposition or reaction at ambient conditions; is a forbidden explosive as defined under 49 CFR 173.51, Class A explosive as defined under 49 CFR 173.53, or Class B explosive as defined under 49 CFR 173.88.

Table 5-2 Characteristics of Hazardous Waste *(continued)*

Characteristic	EPA hazardous waste no.	Properties
Toxicity	D004–D043	Extract from a representative sample contains any of the contaminants listed in Table 1 at a concentration equal to or greater than the respective value provided in Table 1, using the Toxicity Characteristic Leaching Procedure (TCLP; Method 1311) or equivalent method.
Acute hazardous characteristics[a]		Refer to hazardous waste as defined above.
Hazardous waste characteristics		
Human toxicity		Fatal to humans in low doses; in absence of data on human toxicity, studies showing an oval LD_{50} toxicity (rat) of less than 50 mg/kg, inhalation LC_{50} toxicity of less than 2 mg/l, or a dermal LD_{50} toxicity (rabbit) of less than 200 mg/kg or significantly contributes to an increase or irreversible or incapacitating illness.
Toxic waste chemical toxicity		Contains any of the toxic constituents contained in Appendix VII (40 CFR 261)[b]; capable of posing a substantial present or potential hazard when improperly treated, stored, transported, or disposed of, or otherwise managed.

[a] Applies to a solid waste being characterized as a hazardous waste (40 CFR 261.11).

[b] Reflects 418 substances as of July 1, 1991, shown in scientific studies to have toxic, carcinogenic, mutagenic, or teratogenic effects on humans or other life forms.

waste is derived from the use of radioactive materials in medical practices, research laboratories, and industrial processes, notably within the mining industry. In the United States, as much radioactive waste is generated from miscellaneous sources as is generated from the nuclear power industry, although the half-lives of such material are usually restricted to less than a few decades.

Infectious waste comprises a minor category of potentially hazardous waste. This waste type includes (1) any equipment, instruments, utensils, and fomites of a disposal nature from the rooms of patients who are suspected of having or have a communicable disease and therefore must be isolated, (2) laboratory wastes such as pathological specimens, and (3) surgical operating room pathological specimens and disposable fomites. These wastes play no role in the reuse and recycling of contaminated soils.

The Toxicity Characteristics (TC) rule is important because it includes a significant volume of waste previously unregulated in the federal hazardous waste management system. The TC rule added 25 chemicals to the 8 metals and 6 pesticides on the previous Extraction Procedure Toxicity Characteristic list. The rule also replaced the Extraction Procedure (EP) with the Toxicity Characteristic Leaching Procedure (TCLP), which is further discussed under Section 5.6.1. The resultant regulatory levels for the newly added 25 chemicals are determined by the product of a health-related concentration threshold and a dilution/attenuation factor (DAF), which is derived from a newly developed groundwater transport model.

A soil is considered hazardous or toxic if the affected soil exhibits any of the characteristics of a waste (i.e., ignitability, corrosivity, reactivity, or toxicity) or contains a constituent that is a listed waste. The soil can also be considered hazardous should, upon chemical testing, a constituent or several constituents exceed an established maximum contaminant level (or regulatory level). This determination is made using one or a combination of leachability tests as discussed under Section 5.6.

Where no regulatory level exists for the constituent(s) of concern, remedial options are technically or economically limited (i.e., affected soil beneath a major thoroughfare, active and operational facilities, etc.); the affected soil may be deeply seated in low-permeability soil, or the concentrations may not be significantly elevated such that they pose a threat to public health, safety, and welfare or overall groundwater resources. In such circumstances, remedial or corrective action may be limited provided that an environmental risk assessment is performed. An assessment or inventory of health and environmental risks or factors posed by site-specific conditions is an important component of any subsurface characterization per the National Contingency Plan [40 CFR Part 300.430(d)(2)]. Evaluations of site risk factors are performed to demonstrate that no significant adverse effects on human health are anticipated. Such factors include discussion of site geology and hydrogeology, meteorology, ecology, air clarification, waste characteristics, contaminant source, exposure pathways, and Applicable or Relevant and Appropriate Requirements

(ARARs). The ARARs for the site include consideration of federal, state, and local guidelines and cleanup levels.

Overall, there is a general absence of uniform cleanup standards for soil. Reviews of case histories show that site-specific studies are expensive and time-consuming, and produce inconsistent results. The wide range in cleanup levels for soil reflects different risks of exposure and contaminant migration. Multiple methods exist for performing risk assessments due to different ways of interpreting toxicological data and various ways of understanding what conditions constitute acceptable risk. Another factor for varying policies for determining cleanup levels reflects regional differences in environmental deterioration and degree of public involvement.

5.4 ACCEPTANCE CRITERIA

Acceptance criteria typically require the generator (or client) to provide at minimum a description of the candidate soil and appropriate analytical data on such material. For example, in the case of diesel- and/or gasoline-affected soil, total recoverable petroleum hydrocarbons (EPA Method 418.1), total petroleum hydrocarbons (EPA Method 8015 modified), and volatile aromatic hydrocarbons (notably, benzene, toluene, ethylbenzene, and total xylenes using EPA Method 8020) are required. Should metals or other contaminants of concern be a potential issue, then analytical data for these contaminants of concern should also be provided prior to acceptance of the affected soil for reuse and recycling consideration. Should analytical data not be available, then the appropriate tests will need to be performed prior to consideration for processing (Table 5-3). A complete listing of appropriate tests when the contaminant of concern is unknown is presented in Table 5-4.

Prior to consideration of a candidate soil for reuse or recycling, a synopsis of the soil history, site use (i.e., underground storage tank, metal plating facility, etc.), and anticipated contaminant(s) should be made. Such information can alleviate the need for the chemical testing of certain parameters listed in Table 5-3 not anticipated to be present.

5.5 LEACHABILITY

As surface water or groundwater either contacts or passes through various materials, the various constituents that comprise the material will dissolve at some finite rate. The basis of a leachability test in regard to soil, waste, or asphaltic or cementitious end product is that if water is allowed to penetrate a material, certain constituents such as waste will dissolve, and whether untreated, treated, or reused or recycled, exposure to water will result in a rate of dissolution that can be measured. This process is called leaching. Prior to penetration the water is referred to as the leachant, whereas once the water has passed through the material, it is referred to as the leachate. Leachability is the capacity for a material to leach.

Table 5-3 Suggested Chemical Analyses Required Prior to Material Acceptance

Parameter	EPA method
Total petroleum hydrocarbon	418.1
Petroleum hydrocarbon range to C-30	8015 modified with extended hydrocarbon chain
Leachable hydrocarbon content	8015 modified for leachability[a]
Ignitability	1010[b]
Organochlorine compounds (including PCBs)	8080
Volatile organic compounds	8240[c]
Semivolatile organic compounds	8270
Reactivity	9010/9030
Corrosivity	9045
Title 22 metals; total threshold limit concentration	(TTLC)ICAP/AA[d]
Organic, semiorganic, and metals leachability	Toxicity characteristic leaching procedures[e] (TCLP)
Toxicity[f]	96-h Bio Aquatic Assay

[a] While no agency approved laboratory method is available for this test, some certified laboratories have developed a method to determine the leachability of hydrocarbon constituents within the candidate soil matrix. This information is essential to documentation that the reuse and recycling procedure produces a nonleachable product. By way of explanation, the leachability procedure is much the same as the toxicity characteristic leaching procedure (TCLP) as set forth by CCR Title 22 and the Code of Federal Regulations (CFR) Title 40.

[b] EPA Method 1010 for ignitability does not apply to solids. However, the Cal EPA Waste Classification Branch has not decided on a definitive ignitability test for solids; yet, Department of Transport and the California Highway Patrol Hazardous Waste Strike Force requires an ignitability test be performed to determine hazardability. In lieu of a new test method, EPA Method 1010 is performed to comply with conflicting state agency interpretations.

[c] This test, and indeed all tests, are suggestions in order to document constituents that are, or are not, present in the preprocessed affected soil. From the standpoint of multilevel peer and industry review as well as national importance of possible submittal to EPA, this conservative approach would provide substantiation to pilot study veracity.

[d] Presence of regulated metal constituents (if any) and their relative concentrations may be determined by California Code of Regulations (CCR) Title 22 Section 66261.24 criteria, as it is the most conservative of any of the states and even more so than the federal government. All samples are subjected to Total Threshold Limit Concentrations (TTLC) testing. Those metals indicating ten times the Soluble Threshold Limit Concentration (STLC) or ten times TCLP regulatory levels are tested by both STLC and TCLP procedures.

[e] TCLP levels have been established for some volatile and semivolatile organic compounds as well as some elemental metals. Constituents detected by other methods listed in Table 2-2 may also be analyzed by TCLP procedures where possible to provide pre- and post-incorporation comparisons.

[f] Performance of 96-h Bio Aquatic Assays (CCR Title 22, Sec. 66261.24) on the PNAS, as well as the resultant asphalt product, in order to document the relative toxicity (if any) of the preprocessed soil and product as they relate to the process and ingredients utilized to produce the end product is recommended. In addition, conduct of the 96-h Bio Aquatic Assays may serve to document classification of preprocessed PNAS as nonhazardous, which in turn may serve to deregulate the actual remediation site from its current status as a "hazardous waste site" to one of a more benign status [i.e., the constituents may be "listed" hazardous waste per CFR 261 Appendix VII and VIII, but are they in their present state, proven to be hazardous (per Federal Register (FR) January 2, 1992]. Refer to CFR 261, "Materials Containing Constituents Must be Proven Hazardous."

Table 5-4 Summary of Various Leachability Tests

Test procedure	Method no.	Purpose	Material application	Regulatory requirement
Toxicity Characteristics Leaching Procedure (TCLP)[a]	EPA SW-846, Method 1311	Compares toxicity data with regulatory level; includes VOCs; stimulates sanitary or municipal landfill conditions	Stabilized wastes Reused/recycled end-products	RCRA
Extraction Procedure Toxicity (EP Tox) Test[c]	EPA SW-846, Method 1310	Evaluates leachate concentrations; stimulates sanitary or municipal landfill conditions	Stabilized wastes	RCRA
California Waste Extraction Test (Cal Wet)[c]	California Code Title 22, Article 11, pp. 1800.75-1800.82	Provides a more stringent leaching test for metals than TCLP	Hazardous waste classification	California
American Nuclear Society Leach Test (ANSLT)[c]	ANSI/ANS/16.1	Establishes a diffusion coefficient for comparison of S/S-treated waste	Leaching resistance of stabilized wastes	NRC
TCLP "Gage" Modification	53 FR 18792	Adds qualitative evaluation of stability to TCLP test	Stabilized waste Reused/recycled end products	RCRA (Proposed)
Multiple Extraction Procedure (MEP)	EPA SW-846, Method 1320	Evaluates waste leaching under acid rain conditions	Delisting USEPA-listed wastes	None
Synthetic Acid Precipitation Leach Test	EPA SW-846, Method 1312	For waste exposed to acid rain	Hazardous waste characterization	RCRA (Comment)

Monofilled Waste Extraction Procedure (MWEP)	SW-924	For waste disposal in low-velocity saturated zone	Monolith or crushed waste	None
Dynamic Leach Test	WTC, 1991, p. 17	Estimates diffusion coefficient for an S/S-treated waste	Hazardous waste characterization	None
Shake Extraction Test	ASTM D 3987-85	Provides a rapid means of obtaining an aqueous extract	Inorganic compounds	None
Equilibrium Leach Test (ELT)	WTC, 1991, p. 16	Evaluates maximum leachate concentrations	Hazardous waste characterization	None
Sequential Extraction Test (SET)	Bishop, 1986, p. 240	Evaluates buffering capacity with multiple extractions	Hazardous waste characterization	None
Sequential Chemical Extraction (SCE)	WTC, 1991, p. 17	Evaluates bonding nature of metals and organics in the S/S-treated waste	Stabilized wastes	U.S. DOE
Static Leach Test (ambient or high temperature)	MCC-1P, MCC-2P	Evaluates the leach resistance of a bulk specimen in static fluid	Stabilized wastes	None
Agitated Powder Leach Test	MCC-3S	Evaluates the leach resistance of a bulk specimen in agitated fluid	Nuclear waste forms	None
Soxhlet Leach Test	MCC-5S	Evaluates the leach resistance of a bulk specimen in constantly refreshed pure leachant, typically at elevated temperature	Stabilized nuclear waste	None

[a] Regulatory requirement.
[b] RCRA — Resource, Conservation and Recovery Act.
[c] NRC — Nuclear Regulatory Commission.

Modified from Means, J. L., et al., 1995, *The Application of Solidification/Stabilization to Waste Materials,* CRC/Lewis, Boca Raton, FL, 334 pp.

Leachability is typically expressed in concentration of the constituent of concern in the leachate and is compared with its respective concentration in the original material and with various standards, notably drinking water standards. Leachability, however, actually reflects the rate at which a constituent is removed from the material and enters into the environment via the leachate. Despite the fact that we commonly refer to the leachability of a particular constituent as concentration, in the leachate, although neither of the various leachability tests simulates actual conditions or environments to be encountered, they do provide a worst case environment for leaching. Such tests neither reflect leach rates exhibited by natural minerals in nature nor do laboratory leachability test results correlate favorably with field data. In any case, in many instances more than one type of leachability test may be utilized.

Factors affecting leachability can be divided into two groups: those associated with the material being tested and those associated with the test itself. Those factors that are associated with the material being tested include

- pH control
- redox potential control
- chemical reactions
- adsorption
- chemisorption
- passivation
- ion exchange
- diadechy
- reprecipitation
- encapsulation
- alteractions of waste properties

Those factors that are a function of the testing methods include

- Increasing the surface area of the waste via grinding or crushing (this is important since most processes rely in part on chemistry and micro- or macro-encapsulation)
- Variability in the geometry of the extraction vessel, which affects the degree of abrasion the material is subjected to during testing and thus, particle size
- Variable agitation procedures and equipment
- Nature of the leachant
- Ratio of leachant to the waste
- Number of elutions used
- Time of contact
- Temperature
- pH adjustment
- Separation of extract

- Difficulties in both accuracy and reproducibility between the analysis of water and waste residuals and leachates

5.6 LEACHABILITY TESTS

Several leachability tests are available, with the TCLP being the most commonly used. Regardless of which test or tests are used, the appropriateness of certain tests should be confirmed by the lead regulatory agency involved with the project prior to initiation of any reuse or recycling program. There are 16 various tests currently used to assess the leachability of hazardous or radioactive materials and stabilized waste. Although all of these tests provide useful information, many are experimental in nature, and only four are currently actual regulatory requirements, TCLP, EP Tox, Cal WET, and ANSLT. A summary of various leachability tests available is presented in Table 5-4.

In addition to the type of leachability test to be utilized for a particular project, other considerations include contamination concentration reduction in the preprocessed soil due to sample size reduction or in the processed end product due to addition of aggregate and/or binding agent (i.e., emulsion). Another consideration involves size reduction of the produced product for subsequent leachability testing. Size reduction is accomplished either by fragmentation, grinding, or sizing. Size reduction is of significant concern in dealing with testing of the produced product, where preliminary size reduction of the sample is avoided in order to test the potential leachability of the product, not the components or constituents that make up the product.

5.6.1 Toxicity Characteristic Leaching Procedure (TCLP)

TCLP is the most common leaching test used; it is the test most generally accepted by the U.S. EPA for leaching tests and for determining toxicity under RCRA. The TCLP is designed to determine the potential mobility of both organic and inorganic constituents in either liquid, solid, or multi-plastic materials. From a regulatory perspective, the TCLP is used to evaluate the leaching of metals, volatile and semi-volatile organic compounds, and pesticides, among other waste types, from waste categorized under RCRA as hazardous or toxic. TCLP parameters and regulatory MCLs are presented in Table 5-5.

Depending on the alkalinity of the material being tested, particle sizes that have been reduced to less than 9.5 mm are extracted with an acetate buffer solution of pH of 5 or acetic acid solution of pH of 3. The TCLP leachate is, however, poorly buffered. Thus, depending on the initial alkalinity of the waste, the leachate upon contact with the waste may have a pH of 10 or greater.

The solid sample is extracted with an amount of extraction fluid equal to 20 times the weight of the sample. The extraction fluid used in this modified version of the TCLP test is ASTM Type II water. The extraction vessel, a polyethylene bottle, is then placed in a rotary agitator and rotated for 18 h at 30 rpm in a temperature-controlled environment. Following extraction, the

Table 5-5 Toxicity Characteristic Leaching Procedure (TCLP) Parameters and Regulatory Levels

Parameter	Regulatory level (mg/l)	Parameter	Regulatory level (mg/l)
Metals		p-Cresol	200.0[b]
Arsenic	5.0	Cresol	200.0[b]
Barium	100.0	2,4-Dinitrotoluene	0.13[a]
Cadmium	1.0	Hexachlorobenzene	0.13[a]
Chromium	5.0	Hexachlorobutadiene	0.5
Lead	5.0	Hexachloroethane	3.0
Mercury	0.2	Nitrobenzene	2.0
Selenium	1.0	Pentachlorophenol	100.0
Silver	5.0	Pyridine	5.0[a]
		2,4,5-Trichlorophenol	400.0
Volatiles		2,4,6-Trichlorophenol	2.0
Benzene	0.5		
Carbon Tetrachloride	0.5	Organochlorine Pesticides	
Chlorobenzene	100.0	Chlordane	0.03
Chloroform	6.0	Endrin	0.02
1,4-Dichlorobenzene	7.5	Heptachlor (and its epoxide)	0.008
1,2-Dichloroethane	0.5	Lindane	0.4
1,1-Dichloroethylene	0.7	Methoxychlor	10.0
Methyl ethyl ketone	200.0	Toxaphene	0.5
Tetrachloroethylene	0.7		
Trichloroethylene	0.5	Chlorophenoxy Acid	
Vinyl Chloride	0.2	Herbicides	
		2,4-D	
Semivolatiles		2,4,5-TP (Silvex)	
o-Cresol	200.0[b]		
m-Cresol	200.0[b]		

[a] Quantitation limit is greater than regulatory level. Therefore, the quantitation limit is the regulatory level.

[b] If o-, m-, and p-cresol cannot be differentiated, total cresol concentration can be used.

Used for Resource Conservation and Recovery Act (RCRA) regulated hazardous waste. Source is 40 CFR, Part 261.24 and California Code of Regulations, Title 22, Chapter 11, Article 3.

liquid extract is separated from the solid phase by filtration through a 0.6 to 0.8 μm glass fiber filter. Preservatives are not added to samples prior to extraction. Samples to be analyzed for volatiles are routinely refrigerated at 4°C. Samples to be analyzed for metals, however, must be acidified with nitric acid to a pH < 2 after extraction. TCLP extracts are analyzed immediately following extraction. The holding time for organics from collection until extraction is 14 d and from extraction to analysis also 14 d. For inorganics, holding times are 180 d from collection to extraction for all metals, with the exception of mercury, which is 28 d. From extraction to analysis, holding time for inorganics is 180 d, with the exception of mercury, which is 28 d. For quality control and assurance, a blank is included for each batch of 20 samples. A matrix spike and matrix spike duplicate are also included in each such batch. The matrix spike is prepared by post-spike, that being spiked after final filtration but before acidification, when applicable.

The TCLP does not address long-term stability and thus has limitations in regard to the long-term leaching of stabilized waste. However, until another leaching test or combination of tests is mandated by regulation, TCLP remains the most common test for determining the potential leachability of reused and recycled products.

5.6.2 Extraction Procedure Toxicity (EP Tox) Test

The EP Tox leachability test is the precursor of the TCLP. Although very similar to the TCLP, only one concentration of acetic acid solution of pH of 5 is used. In addition, in lieu of a liquid-to-solid ratio of 20:1, a ratio of 16:1 is used, which may increase as additional acid solution is added during the 24-h duration of the test to adjust the pH as needed.

At pH of five, the EP Tox test results compare favorably with those attained via the TCLP. However, such results may significantly differ at a lower pH (i.e., pH of 3). The measurement of pH in the extract during conduct of either the EP Tox or TCLP can assist in evaluating whether the leaching of a particular contaminant is pH-dependent. In addition, the EP Tox test is not applicable to evaluate volatile organics.

5.6.3 California Waste Extraction Test (Cal WET)

Certain states, such as California, maintain double standards and have set forth further regulations to determine if a material is to be classified as hazardous or nonhazardous. The decision about how the material is classified is based in part on the analytical values obtained in the lab. The regulations provide two different tables of maximum allowable threshold concentration values, the total threshold limit concentration (TTLC), and the soluble threshold limit concentration (STLC). Parameters and regulatory levels for TTLC and STLC are presented in Table 5-6.

TTLC values apply to samples that have undergone a very harsh extraction protocol that is performed using concentrated nitric acid and hydrogen peroxide or hydrochloric acid (EPA Method 3050, contained in SW-846). The values obtained from this procedure can be, for all practical purposes, interpreted as the total quantity of a particular element in that sample. The second value is

Table 5-6 Total Threshold Limit Concentration (TTLC) and Soluble Threshold Limit Concentration (STLC) Parameters and Regulatory Levels[a]

Parameter	Regulatory level	
	TTLC[b] (mg/kg)	STLC[c] (mg/l)
Inorganic compounds		
Antimony	500	15
Arsenic	500	5.0
Barium	10,000[d]	100
Beryllium	75	0.75
Cadmium	100	1.0
Chromium	2,500	5[e]
Cobalt	8,000	80
Copper	2,500	25
Lead	1,000	5
Mercury	20	0.2
Molybdenum	3,500	350
Nickel	2,000	20
Selenium	100	1
Silver	500	5
Thallium	700	7
Vanadium	2,400	24
Zinc	5,000	250
Chromium (VI)	500	5
Fluoride Salts	18,000	180
Asbestos	1%	—
Organochlorine pesticides and polychlorinated biphenyls		
Aldrin	1.4	0.14
Chlordane	2.5	0.25
DDT/DDE/DDD	1.0	0.1
Dieldrin	8.0	0.8
Endrin	0.2	0.02
Heptachlor	4.7	0.47
Kepone	21	2.1
Lindane	4.0	0.4
Methoxychlor	100	10
Mirex	21	2.1
PCBs	50	5.0
Toxaphene	5	0.5
Volatiles		
Trichloroethylene	2.040	204

Table 5-6 Total Threshold Limit Concentration (TTLC) and Soluble Threshold Limit Concentration (STLC) Parameters and Regulatory Levels[a] *(continued)*

Parameter	Regulatory level TTLC[b] (mg/kg)	STLC[c] (mg/l)
Chlorophenoxy acid herbicides		
2,4-Dichlorophenoxyacetic acid (2,4-D)	100	10
2,4,5-Trichlorophenoxypropionic acid (Silvex)	10	1.0
Semivolatiles		
Pentachlorophenol	17	1.7
Miscellaneous		
Dioxin (2,3,7,8-TCDD)	0.01	0.001
Organic lead	13	—

[a] Used for California regulated hazardous waste. Source is California Code of Regulations, Title 22, Chapter 11, Article 3.

[b] If a substance in a waste equals or exceeds the TTLC level, it is considered a hazardous waste.

[c] If a substance is 10 times (by rule of thumb) the STLC value as found on the TTLC, the WET should be used. If any substance in the waste so analyzed equals or exceeds the STLC value, it is considered a hazardous toxic waste.

[d] Excludes barium sulfate.

[e] If soluble chromium, as determined by TCLP test, is less than 5 mg/l and soluble chromium as determined by the STLC test equals or exceed 560 mg/l and waste is not otherwise identified as RCRA hazardous waste, then the waste is non-RCRA hazardous.

known as the STLC and applies to samples that have gone through a very mild leaching procedure with citric acid. This method is also referred to as the WET extraction. It supposedly replicates *in vitro* the leaching process that may potentially occur in a sanitary landfill environment. As water migrates through domestic garbage, picking up natural acids such as citric acid, it then continues on its migratory pathway through the waste in question, leaching into solution elements that could possibly travel down to the water table. This *in vitro* process consists of tumbling the sample for 18 h in a buffered solution of pH 5 of citric acid. The ratio of sample to leaching solution is 1:10. That means that if you would multiply the result of this WET extraction by 10, you would get a rough total value for this sample, if and only if the sample's analyte in question is completely soluble in water, which is a possibility that conservatively cannot be discarded.

Based on this 1:10 TTLC/STLC relationship, if any of the total values for a sample exceeds a value 10 times the STLC limit, it may possibly exceed the STLC values when the WET extraction test is performed. Thus, it is very practical and cost effective to schedule a set of samples to be analyzed first for totals, and if any of the values exceeds 10 times the regulatory STLC limit, then conduct of the WET extraction for only the element or elements in question is performed.

The actual test utilizes a sodium citrate buffer, with a liquid-to-solids ratio of 10:1. Testing duration is 48 h. The Cal WET is more aggressive in comparison to the TCLP or EP Tox. As a result, a category of waste specific to California has been developed. If a waste passes the TCLP but fails the Cal WET, then the material is categorized as a "California-only" hazardous waste. Furthermore, if a material fails both the TCLP and Cal WET, then both California and USEPA criteria must be met.

5.6.4 American Nuclear Society Leach Test (ANSI/ANS/16.1)

The ANSI/ANS/16.1 leach test is performed on treated waste or reuse and recycled end products. This test is utilized to develop a figure-of-merit, which is used to compare the leaching resistance of the material being tested. Recorded in terms of cumulative fraction leached relative to the total mass of the material, the results are used to derive an effective diffusion coefficient and leachability index (or figure-of-merit).

The objective of the ANSI/ANS/16.1 test is to arrive at a contaminant release rate. This test is conducted over a period of 90 d. The leachant is typically distilled water, but occasionally other liquids such as simulated groundwater may also be used.

5.6.5 Toxicity Characteristic Leaching Procedure "Gage" Modification (TCLP Modified)

As with the TCLP test, the TCLP modified was designed to quantify the mobility of both organic and inorganic contaminants in solid samples. The TCLP modified does not require samples to be passed through a 9.5 minus screen or to meet certain surface area requirements prior to conduct of the leachability test. This is important in regard to reuse and recycling technologies since such a procedure does not treat the end product but rather the components that comprise the asphaltic or cementitious end product. This test also differs from the standard TCLP in that it addresses treated waste or waste in produced solidified asphaltic or cementitious end products, which theoretically would withstand environmental stresses anticipated to be encountered in a landfill. The objective is to maintain the structural integrity of the produced product. The product sample is thus tumbled in a gage and left more or less intact, with no size reduction required outside of obtaining a product sample small enough to fit into the sample gage. Since the TCLP modified is not regulation driven, such a procedure should be approved by the lead regulatory agency prior to conduct of such tests.

5.6.6 Dynamic Leach Test (DLT)

A modified version of the ANSI/ANS/16.1 test is the DLT. The DLT is used to determine a diffusion coefficient, which is then used to predict long-term leaching performance.

Analytically, the renewal frequency of the leaching solution is adjusted to ensure that equilibrium has not been reached. The leaching volume-to-solid ratio is also adjusted to ensure that contamination can be detected. Such adjustments rely on an estimated diffusion coefficient and result from batch tests as performed under the ELT, as discussed under Section 5.6.11.

5.6.7 Multiple Extraction Procedure (MEP)

The MEP is used to simulate leaching in an improperly designed landfill where the waste could potentially come into contact with large volumes of acidic leachate. This test has also been used for the delisting of U.S. EPA-listed wastes.

Similar analytically to EP Tox, the initial extraction is performed with acetic acid but is then followed by a minimum of eight extractions performed with a synthetic acid rain solution, notably, a sulfuric/nitric acid solution that has been adjusted to a pH of 3.

Most metals solubility increases with decreasing pH. The MEP is thus superior to the TCLP in assessing leaching behavior of contaminants, notably metals, as a function of decreasing pH. This is because the MEP gradually removes excess alkalinity in the waste material with time as a result of the numerous extractions performed.

5.6.8 Synthetic Acid Precipitation Leach Test (SAPLT)

The SAPLT simulates acid rain as opposed to leachate in a landfill, as do the TCLP and EP Tox tests. Analytically similar to the TCLP, the initial liquid-solid separation phase is, however, excluded, and replaced with a dilute nitric acid/sulfuric acid mixture.

5.6.9 Monofilled Waste Extraction Procedure (MWEP)

The objectives of the MWEP is to derive leachate compositions in mono-filled disposal sites or leachate for evaluating the compatibility of various lining materials with the leachate. This test consists of multiple extractions of a monolith or of crushed waste with distilled and/or ionized water.

The sample is crushed to less than 9.5 mm or left intact, providing it passes the USEPA SW-846 Structural Integrity Test. The liquid-to-solid ratio is 10:1. Extraction of samples with water is performed four times at 18 h per extraction.

5.6.10 Shake Extraction Test (SET)

The SET is a rapid means of retrieving an aqueous extract. This test is not intended to simulate site-specific leaching conditions. During this test, the leachate of a solid waste is extracted with Type IV reagent water in a rotary agitator for 18 h. This test is only applicable to inorganic compounds.

5.6.11 Equilibrium Leach Test (ELT)

The ELT is utilized to determine equilibrium leachate concentrations under mild leaching conditions. Static leaching is performed with distilled water on a crushed sample of less than or equal to 150 μm. This smaller size relative to the TCLP and EP Tox results in greater contact surface area and reduced time required for equilibrium to be reached. During conduct of the ELT, the leachate is added once at a liquid-to-solid ratio of 4:1. The sample is then agitated for 7 d.

5.6.12 Sequential Extraction Test (SET)

The SET is utilized to evaluate the waste buffering capacity and alkalinity of cement-based stabilized and solidified treated waste or reused and recycled end products. Fifteen extractions of one sample of crushed waste with particle sizes ranging between 2.0 and 9.5 mm is performed with a 0.04 *M* acetic acid solution. Each extraction is conducted on what is referred to as a shaker table for 24 h with the same type of extraction solution. A liquid-to-solid ratio of 50:1 is used. A 2 meq/g of acid is added to the crushed waste, the pH is measured, and the leaching solution is filtered with each extraction. After the fifteenth extraction, the residual solids are digested with three more extractions with more concentrated acid solutions used.

5.6.13 Sequential Chemical Extraction (SCE)

The SCE is utilized to evaluate the nature of and banding strength of metals and organics in stabilized and solidified treated waste and/or reused and recycled end products. Originally developed for sediments, the SCE was later adapted to evaluate inorganic waste constituents incorporated within a stabilized matrix. Although the SCE is similar to the SET in that it involves sequential extractions, the leaching solution, sequentially increases in acidity from neutral to very acidic. In addition, the sample particle size is smaller, being less than 45 μm.

5.6.14 Static Leach Test Method (SLTM)

The SLTM, as developed by the Material Characterization Center (MCC) at Pacific Northwest Laboratory under the auspices of the U.S. DOE, is utilized to evaluate the chemical durability and leach resistance of stabilized and solidified treated nuclear waste. Representative monolithic samples of material of a known geometric surface area are immersed (but not agitated) in a referenced leachant at a specified temperature. Immersion can vary from three days to years. Temperatures can range from ambient to high temperatures (i.e., 40 to 190°C).

5.6.15 Agitated Powder Leach Test (APLT)

The APLT is utilized to evaluate the chemical durability of nuclear waste forms. Also developed under the MCC, this test uses representative powdered waste samples, which are immersed in a referenced leachant at a constant ratio of leachant volume to specimen mass of 10 mx/g. Agitation is conducted by constant rolling of the specimen holder at temperatures ranging from 40 to 190°C.

5.6.16 Soxhlet Leach Test (SLT)

The SLT was also developed by the MCC to evaluate the chemical durability of stabilized and solidified treated nuclear waste. Originally developed primarily for glass and ceramic waste forms, the SLT is applicable to monolithic treated stabilized and solidified waste, and the individual components of macro scale physical composite treated wastes.

The SLT is conducted with monolithic samples of a known geometric surface that are suspended in a continuously flowing stream of redistilled water. Test temperatures are precisely determined by the barometric pressure in the laboratory, but usually is near 100°C and normalized mass losses are measured.

BIBLIOGRAPHY

Batchelor, B. and Wu, K., 1992, Effects of equilibrium chemistry on leaching of contaminants from stabilized/solidified wastes, in *Chemistry and Microstructure of Solidified Waste Forms* (Edited by R. D. Spence), CRC/Lewis, Boca Raton, FL, pp. 243–259.

California Environmental Protection Agency, 1992, Commercial Hazardous Waste Facilities for Recycling/Recovery, Treatment, Disposal (Draft), Document No. 902, 97 pp.

Cocke, D. L. and Mollah, M. Y. A., 1992, *The Chemistry and Leaching Mechanisms of Hazardous Substances in Cementitious Solidification/Stabilization Systems* (Edited by R. D. Spence), CRC/Lewis, Boca Raton, FL, pp. 187–242.

Conner, J. R., 1990, *Chemical Fixation and Solidification of Hazardous Waste*, Van Nostrand Reinhold, New York, 692 pp.

Means, J. L., et al., 1995, *The Application of Solidification/Stabilization to Waste Materials*, CRC/Lewis, Boca Raton, FL, 334 pp.

United States Environmental Protection Agency, 1990, *Federal Register*, Thursday, March 29, 1990, 40 CFR Part 261.

6 ENGINEERING CONSIDERATIONS

6.1 INTRODUCTION

Those parameters important in addressing engineering concerns associated with the incorporation of contaminated soil into asphalt can be divided into two groups, parameters and tests applicable to preprocessed soil and those applicable to processed asphalt. These parameters and tests for preprocessed soil include microscopic examination, sieve analysis, sand equivalent test, maximum density and moisture content, specific gravity, and R-value or resistance. Those tests that are applicable to the processed asphaltic end product consist primarily of Marshall stability and development of mix designs. Requirements of pilot or bench tests prior to full-scale implementation of a contaminated soil reuse and recycling program are also discussed. Not specifically addressed is workability. Workability is one of the most difficult qualities to evaluate by conventional test methods. Workability can be defined as the ease and speed with which the asphaltic material can be applied and compacted while achieving the specification density and finished surface desired (i.e., absence of deformities and segregation). One must rely more on experience, which must always be taken into account, when determining workability. Presented in this chapter is discussion of the various parameters and engineering tests associated with preprocessed soil and processed asphalt.

Presented in this chapter is discussion of engineering tests typically performed on preprocessed soil and processed asphalt, and mix designs in regard to uses and criteria. The objective and various components of pilot testing are also discussed.

6.2 ENGINEERING TESTS FOR PREPROCESSED SOIL

Conventional engineering tests performed on preprocessed soil samples include microscopic examination, sieve or particle size distribution analysis, sand equivalent, maximum density and moisture content, specific gravity, and

R-value or resistance. A summary of these tests, their purposes, and testing methods is presented in Table 6-1 and further discussed below.

6.2.1 Microscopic and Petrographic Examination

Microscopic examination is performed to identify the physical characteristics of the material being incorporated. Parameters of importance include shape, soundness, and aggregate identification. Petrographic examination is occasionally used to determine aggregate identification and source and in evaluating porosity and degree of encapsulation (Figure 6-1). Petrographic analysis is performed utilizing a polarizing microscope, which allows examination of a 0.03-mm thick sample under plain and polarized light.

6.2.2 Sieve Size Analysis

Sieve size analysis is useful in determining the individual size components or particle size distribution of fine and coarse aggregate or of the soil matrix (i.e., percentage of rock, gravel, sand, and fines in total volume). Particle size distribution is also performed as discussed in Chapter 4 to evaluate whether the contaminants of concern are restricted to a particular size fraction, which in the course of testing may be segregated in the field. Other specific test methods also exist specifically for fine- and coarse-grained mineral aggregate-type material finer than No. 200 sieve (75 μm) by washing. Classification and nomenclature used for both ASTM and AASHTO classification schemes are presented in Table 6-2.

Gravelly soil will normally contain at least 15% gravel-sized materials. Sandy soil will contain more than 50% sand- and gravel-sized fractions. Silty soil will contain from 40% to as much as 100% silt-sized fractions, whereas clay can contain as low as 30% or as high as 100% clay and colloids.

6.2.3 Sand Equivalent Test

Sand equivalent testing is performed to determine the expansiveness of the soil, which has a bearing on mix design formulation and ratios. The term expresses the concept that most aggregate, whether coarse or fine, is a mixture of desirable coarse particles and generally undesirable clay or plastic fines and dust. This test is a rapid field correlation test, which indicates the relative proportions of clay-like or plastic fines and dust in granular soil and fine aggregate (that which passes the No. 4 (4.75 mm) sieve under standard conditions).

6.2.4 Maximum Density and Moisture Content

Maximum density is performed to determine the weight per cubic foot and compactibility. Moisture content is used in conjunction with the maximum

Table 6-1 Summary of Engineering Tests

Phase	Engineering test/parameter	Purpose
Preprocessed soil	Microscopic examination	Characterize aggregate and waste
	Sieve size analysis (particle size distribution)	Determine size components Determine size fraction Determine contaminant distribution
	Sand equivalent test	Determine expansiveness of soil Formulate mix design
	Maximum density	Determine compactability Determine weight per cubic foot Formulate mix design
	Moisture content	Formulate mix design
	Liquid limit	Determine soil water content from plastic to liquid state
	Plasticity	Determine plastic state of soil
	Specific gravity	Determine mass of given volume
	R-value or resistance	Evaluate resistance of asphalt pavement under specific loading Determine pavement thickness requirements to minimize or avoid plastic deformation
Processed asphalt	Marshall stability	Determine anticipated stability Determine anticipated strength Document conformity to industry standards Evaluate mix design to a product to be insensitive to moisture effects Evaluate suitability of mix design for anticipated end use
	Mix design	Determine overall performance

Table 6-1 Summary of Engineering Tests *(continued)*

Phase	Engineering test/parameter	Purpose
	Aggregate gradation selection	Evaluate packing and compaction characteristics of aggregate (i.e., shape, texture, and gradation) for anticipated mix design
	Compaction hammer	Evaluate differences between simulated end point traffic density (i.e., ultimate compaction density to be achieved under traffic)
	Optimum asphalt content	Determine optimum asphalt or emulsion content for anticipated mix design
	Stabilometer value	Determine optimum asphalt content
		Evaluate aggregate internal friction (i.e., crushed content, angularity, surface roughness, and gradation)
	Flow value	Evaluate anticipated surface coarse mix
	Air voids	Evaluate acceptable mix design under assumed traffic loads
	Mineral aggregate voids	Evaluate sufficient asphalt cement content for durability
	Dust-asphalt ratio	Determine need for dust control

density and sand equivalent tests to formulate mix designs and to provide a baseline for the assessment of maximum moisture content during processing. Excessive surface moisture can cause problems during mixing, compaction, and curing. Adequate moisture is required during mixing when using slow-setting emulsions (i.e., SS-1 and SS-1h) and anionic grades of medium-setting (i.e., MS-1, MS-2, and MS-2h) emulsions. HFMS grades and CMS-2 and CMS-2h emulsions contain a quantity of petroleum distillates and perform better with dry aggregates.

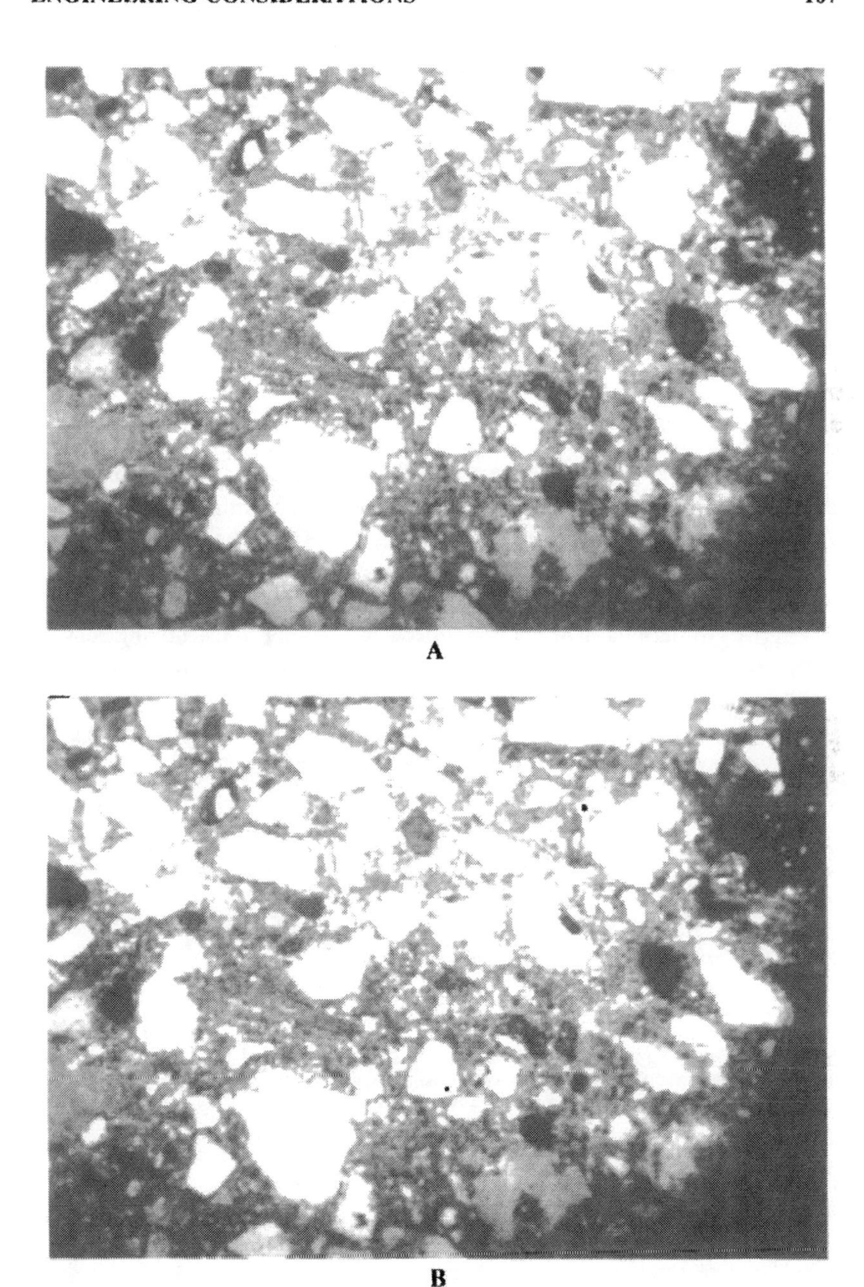

A

B

Figure 6-1. Photomicrographs of petrographic thin sections of processed CMA under plain (A) and polarized (B) light, showing full encapsulation with negligible pore space and connectivity once compacted (3.5 magnification).

Table 6-2 Nomenclature for Certain Grain Size Limits

ASTM D422		AASHTO T88	
Category	**Size interval**	**Category**	**Size interval**
Gravel	75–4.75 mm	Particles	>2.0
Coarse sand	4.75–2	Coarse sand	2.0–0.425
Medium sand	2–0.425		
Fine sand	0.425–0.075	Fine sand	0.425–0.075
Silt	0.075–0.005	Silt	0.075–0.002
Clay	<0.005	Clay	<0.002
Colloids	<0.001	Colloids	<0.001

Since mixing properties are closely related to the density of the compacted specimens, it is necessary to optimize the water content at compaction to maximize the desired mixture properties. Given the maximum density and optimum moisture content, preferred compaction results can be ascertained. Some combinations of aggregate and asphalt emulsion are not significantly affected by variations in water content during mixing; thus, mixing may be sufficient at or above the optimum water content as determined for compaction.

Conventional test methods exist to determine the relationship between moisture content and density of a soil sample when compacted in a given mold of a given size. Several methods are available, depending on grain size passing the No. 4 (4.75-mm) sieve and ¾-in. (19.9-mm) sieve.

6.2.5 Liquid Limit and Plasticity Index

Testing for the liquid limit of a soil is useful to determine the water content at which the soil passes from a plastic state to a liquid state. The lowest water content at which the soil remains plastic is referred to as the plastic limit. The plasticity limit is the range in water content of a soil expressed as the percentage of the mass of an oven-dried sample, within which the material is in a plastic state. The plasticity index can also be viewed as the numerical difference between the liquid limit and the plastic limit of a soil.

6.2.6 Specific Gravity

Specific gravity is the ratio of a mass in air of a given volume of soil or aggregate material at a stated temperature to the mass of air. Of measures using a pycnometer, several methods exist, depending upon grain size: greater than No. 4 (4.75-mm) sieve, less than No. 4 sieve, less than No. 10 (2.00-mm) sieve, etc.

6.2.7 R-Value or Resistance

R-value or resistivity testing utilizes the same apparatus as is used for Marshall stability testing, which is discussed later in this chapter, to evaluate the resistance of the asphaltic product under specific loading for bases, subbases, and subgrades as it relates to pavement thickness design. The resistance to plastic flow of a cylindrical specimen of bituminous paving mixture or cores of asphaltic end product loaded on a lateral surface. R-value testing is used during formulation of mix designs, also discussed later in this chapter; density and voids properties may also be determined.

This test determines the thickness of cover or structural section required to prevent plastic deformation of the soil under imposed wheel loads. The expansion pressure test determines the thickness or weight of cover required to maintain the compaction of the soil. R-values are determined from the moisture content and density at which these two thicknesses are equal. With granular nonexpansive soil, the design R-value is determined for a density considered equivalent to that which will be obtained by normal construction compaction as obtained from exudation data (i.e., compressive stress necessary to exude water from a compacted specimen).

A resistance (R) value test determines or measures the stability or resistance to plastic deformation of compacted materials by means of a Hveem stabilometer and can be used to predict performance of treated and untreated soil.

6.3 ENGINEERING TESTS FOR PROCESSED ASPHALT

Conventional engineering tests typically performed on processed asphaltic end products consist primarily of Marshall stability testing, as is discussed below.

6.3.1 Marshall Stability

Stability tests and mix design methods in general are performed using the Marshall Method and the Hveem Method, the former being more commonly used and compatible with field use. Marshall stability tests are performed to evaluate the anticipated stability of the asphaltic product and to document its conformity to industry standards and specifications. Specifically, Marshall stability tests conducted on product samples can be used to determine whether the produced product will (1) maintain required strength and stability to withstand repeated load applications (compression and flexural) without excessive permanent deformation or fatigue cracking and (2) render the ultimate mixture sufficiently insensitive to moisture effects. For CMA paving mixtures, a minimum value of 2224 pounds at 22.2°C (72°F) is required.

6.4 MIX DESIGNS

Two mix design methods are in general use throughout the United States, the Marshall Method and the Hveem Method. Both of these methods were developed to relate specific measurable mix characteristics and properties to mix performance in a pavement structure. The Marshall Method, which has been in use for over 40 yr, is preferred over the Hveem Method for several reasons. Most important, the Marshall Method is more readily adaptable to varying plant process controls and final mix acceptance and more suitable for preliminary mix design purposes. In addition, since the Hveem Method relies more on recipe formulations, which provide relatively less assurance in regard to mix quality and performance, the Marshall Method is preferred for the measurement of mix properties.

6.4.1 Mix Design Uses

Mix design development employs the Marshall Method as previously discussed and is performed to assure that the end result will be a load-bearing asphaltic product suitable for its intended end use. Development of mix designs also assures that the end product will effectively encapsulate, fixate, and stabilize the constituents of concern into a low permeability, nonleachable product. Mix designs are typically performed on a trial basis prior to full-scale production. Logically based on volumetric considerations, varying amounts of contaminated soil, aggregate, and emulsion are mixed and then subsequently tested for stability and durability.

When considering the potential or intended use of the end product, the mix design requirements for dust abatement or tank containment area will differ significantly from that required for a Class A parking lot, where the latter may have need for three specific mix designs to fulfill the specifications for the entire road section (i.e., sub-base, base, and finished pavement).

Specifications for HMA and CMA differ significantly. HMA maintains a standard specification that is uniformly accepted. For example, a freeway constructed of HMA in California is composed of the same proportion of constituents as an expressway constructed in New York. With CMA, however, there are specifications developed by ASTM, the Asphalt Institute, among other state and local agencies and groups, to fit almost every application where asphalt is needed.

Where no HMA plants are available, notably in the southwest, CMA is commonly used. Arizona, New Mexico, and Texas have utilized CMA for federal highway projects, including Interstates 10, 80, and 40 highway systems. State and county road departments have been using CMA for decades. The "macadam" roads of Washington, D.C. date back to the 1790s and 1800s. Many roads throughout southern California are also constructed of CMA.

Southern California, for example, has a mix design for CMA known as Desert Mix. Desert Mix is composed of road base, native material, and asphalt emulsion. San Diego, in fact, utilizes a CMA mix design that calls for no imported aggregate, but rather uses solely native materials.

6.4.2 Mix Design Criteria

Mix design criteria reflect requirements essential for obtaining a desired degree of end product performance. Performance can be affected by such factors as pavement thickness design, asphalt and aggregate material selection and testing, mix production application (i.e., lay-down), and traffic and environmental conditions. Criteria vary greatly from one use to the next; thus, differences in minimum and/or maximum values are generally the rule. Mix design criteria currently in use include aggregate gradation selection, Marshall compaction hammer, optimum asphalt content, Marshall stabilometer valve, Marshall flow valve, air voids, voids in mineral aggregate, voids in mineral aggregate with asphalt, and dust-to-asphalt ratio.

6.4.2.1 Aggregate Gradation Selection

A wide range of aggregate gradation is typically allowed. Mix designs are commonly based on volumetric considerations and packing or compaction characteristics of the aggregate, with the most important factors being aggregate shape, texture, and gradation. Less importance can be placed on those factors when Marshall design criteria are implemented as a means of mix acceptance. This is because these factors can be evaluated in terms of their influence on mix characteristics with Marshall design criteria, which is really what is important.

6.4.2.2 Compaction Hammer

An uncertainty that occurs during mix design formulation is assessing and anticipating differences between simulated end point traffic density (i.e., ultimate compaction density, which will be achieved under traffic) and density of a compacted mix using the Marshall Method. This relationship is one of the most important elements in developing mix designs. Regardless of the combination of ingredients to produce a mixture, the objective is to produce an asphaltic product such as pavement that is dense enough to prevent the adverse effects of air and water intrusion into the mix. Yet, the product must not be so dense that hydrostatic inability occurs, in which the road is no longer carried by the aggregate matrix but rather by the asphalt cement phase. The number of compaction blows on the Marshall hammer range from 75 to 35 blows per side, or commonly 50 blows per side. However, once a compaction level is selected commensurate with the anticipated traffic diversity to the best of one's ability, and an initial mix gradation is determined, mostly by arbitrary requirements, only the optimum asphalt content needs to be determined for the mixture.

6.4.2.3 Optimum Asphalt Content

The optimum asphalt content is expressed in weight percent. Optimum asphalt content can be determined by averaging the asphalt content values at optimum Marshall stability, optimum mix density, and 4% air voids, and then satisfying the average volume mix design criteria requirement for air voids, Marshall stability, Marshall flow, voids in mineral aggregate, and voids in mineral aggregate filler, if used.

Achieving the asphalt emulsion coating of 100% common to HMA mixes is desirable but not required. This is because sufficient asphalt to achieve 100% coating may result in an excessively high asphalt content.

The estimated percent of asphalt emulsion required for densely graded mixes can be calculated as follows:

$$P = (0.05A + 0.1B + 0.5C) \times (0.7)$$

where P = weight of asphalt emulsion, based on weight of graded mineral aggregate (in percent), A = mineral aggregate retained on 2.36-mm (No. 8) sieve (in percent), B = mineral aggregate passing 2.36-mm (No. 8) sieve, and retained on 75-μm (No. 200) sieve (in percent), and C = mineral aggregate passing 75-μm (No. 200) sieve (in percent). All percentages are typically expressed in whole numbers. When dealing with MC and SC asphalt emulsion requirements, this equation is modified as follows:

$$P = 0.02A + 0.07B + 0.15C + 0.20D$$

where P = asphalt material by weight of dry aggregate (in percent), A = mineral aggregate retained on 300-μm (No. 50) sieve (in percent), B = mineral aggregate passing 300-μm (No. 50) sieve and retained on 150-μm (No. 100) sieve (in percent), C = mineral aggregate passing 150-μm (No. 100) sieve, and retained on 75-μm (No. 200) sieve (in percent), and D = mineral aggregate passing the 75-μm (No. 200) sieve (in percent).

6.4.2.4 Stabilometer Value

Marshall stabilometer values are used by some to select optimum asphalt content, although such values are more important as an evaluation of an aggregate interval friction (i.e., crushed content, angularity, surface roughness, and gradation). Minimum values can range from 500 to 4000 lb, depending on the jurisdictional authority. Maximum values range from 3000 to 3500 lb. Mixes having stabilometer values ranging between 500 to 4000 are usually strong enough to suit any traffic or weight conditions, provided adequate pavement thickness is maintained. Values of 500 to 1000 lb are typically sufficient for traffic, although they can still have a tendency for exhibiting tenderness, shoving, and abrasion loss or raveling. Between 1000 and 35,000 lb, no specific meaning is applied. Above 3500 lb, however, are atypical values that may in fact be detrimental to the performance of the mix. Differences in values between mixes of less than 500 lb do not have much significance, although stringent specifications may not permit such large differences in regard to minimum requirements (i.e., no less than 1800 lb). In summary, the Marshall stabilometer value should be less emphasized as a requirement in mix design and not used to select mix asphalt content.

6.4.2.5 Flow Value

Marshall flow value includes maximum and minimum values. Maximum values range from 16 to 20, with the minimum values ranging from 6 to 8. Maximum values are important in regard to surface coarse mix. Above 20 is unacceptable, whereas heavy traffic usually requires 18 or as low as 16. The minimum value is often ignored, since anything less than 6 reflects a failure of some other criterion such as voids within mineral aggregate.

6.4.2.6 Air Voids

Air voids are the most important parameter in mix design. The value for air voids is typically accepted at 4%, and in some cases as low as 3%. This accepted value is based on a maximum mix density value via the Rice test and is corrected for absorbed water when porous aggregate is used. It is established at a compaction level, which simulates the ultimate mix density (or air void level) anticipated to occur under assumed traffic loads.

6.4.2.7 Voids in Mineral Aggregate

Voids within the mineral aggregate are the second most important parameter in mix designs. Too low a value would not allow enough space in the aggregate matrix, along with the 4% air void requirement, for sufficient asphalt cement for durability. Too high a value would not be cost effective considering the cost for asphalt cement. Minimum values range from 15 to 16% for surface mixes, depending on nominal aggregate sizes. Maximum values are not needed, as they are solely cost related, since the durability of an asphalt mixture is related to the unabsorbed asphalt content of the mixture. Thus, only a minimum parameter for voids in mineral aggregate, based on volumetric measurements, ensures sufficient asphalt cement in a mixture independent of the aggregate's weight and specific gravity.

6.4.2.8 Dust-to-Asphalt Ratio

The dust-to-asphalt ratio is somewhat important with baghouse-type emission control devices and asphalt pavement recycling. To control the cost for asphalt cement, dust should be controlled, with excess dust more of a concern than too little dust or material passing 200 mesh material. A dust-to-asphalt ratio of 1 is desired in plant produced mixes, whereas this is of lesser importance with mobile plants.

6.5 PILOT TESTING

The objective of the pilot test for implementation of reuse and recycling strategies is one of formulating a worst case scenario by incorporating the maximum amount of contaminated soil and the highest concentrations available, while taking into account field variations. Then a determination is made of whether the resultant product can be demonstrated as nonhazardous and nonleachable, while concurrently attaining the engineering requirements dictated by the end use. From a generator's or client's perspective, the goal is to incorporate the maximum percentage of contaminated soil in total volume possible to reduce overall costs. Thus, all three concerns, environmental, engineering, and economic, must be addressed during conduct of a pilot study.

Pertinent to soil reuse and recycling projects, the specific objectives of conducting a pilot or bench study are to

- Determine whether a nonhazardous, regulatory agency-exempt, viable commercial product can be produced;
- Determine through particle size distribution (PSD) analysis whether contaminants of concern predominate within certain sizes of the soil matrix and assess feasibility of on-site mechanical separation;
- Determine the need for volatile organic vapor mitigation via formulation of surfactants, penetrants, and vapor suppressants to control volatile organic emissions;
- Determine appropriate mix design formulations with sufficient Marshall Stability indices (strength and durability) to accommodate the end use requirements;
- Determine through formulation of various cold-mix designs the maximum and therefore most cost-effective ratios of contaminated soil, aggregate, and emulsion for production of various grades of CMA;
- Estimate quantities of produced product for intended use(s);
- Assure incorporation of contaminated soil into an asphaltic product that will effectively exempt contaminated soil, process, and end product from regulation as a hazardous waste;
- Confirm that the incorporation of contaminated soil will be cost effective, time efficient, and environmentally sound.

The tasks associated with pilot or bench tests include sample selection, VOC mitigation testing, PSD analysis, chemical and engineering analysis for

Table 6-3 Summary of Tasks Associated with Pilot Testing

Task no.	Description	Primary tests[a]
1.0	Representative Sample Selection	
2.0	Volatile Organic Compound Mitigation Test	
3.0	Particle Size Distribution Analysis	Particle Size Distribution
4.0	Chemical Analysis for Soil	TCLP; STLC; TTLC
5.0	Engineering Analysis for Soil	Sieve analysis; sand equivalent test; maximum density; moisture content; R-value; resistance; Marshall stability
6.0	Emulsion Selection	
7.0	Mix Design Formulations	Marshall stability
8.0	Asphaltic Product Sample Production	
9.0	Chemical Analysis for Product[a]	TCLP; STLC; TTLC
10.0	Engineering Analysis for Product	Marshall stability
11.0	Regulatory Compliance Review	
12.0	Cost-Benefit Analysis	

[a] Same tests as performed on preprocessed contaminated soil.

contaminated soil samples, formulation of mix designs, production of asphaltic samples, chemical and engineering tests analysis for product samples, regulatory compliance, and cost benefit analysis (Table 6-3). A schematic flowchart illustrating the various sequences of tasks performed during conduct of a pilot test is shown in Figure 6-2.

Every effort should be made to obtain representative samples. From an environmental perspective, it is important to obtain worst-case samples (i.e., those with the highest concentrations and reflecting all potential contaminants of concern). From an engineering perspective, the presence of coarse aggregate and debris, which may need to be segregated and removed, and representation of all soil types are important. Field samples may be obtained in a grid-like manner or at statistically random locations in accordance with conventional U.S. EPA guidelines. A minimum of 25 lb of soil is required per sample. Prior to conduct of field trials or full-scale field operations, representative soil samples can be retrieved from the field and screened for VOCs, whether derived from hydrocarbon-impacted soil or from decaying vegetation. Should mitigation and control of VOC emissions be warranted, identification of the specific constituents and maximum concentrations to be encountered and formulation of the appropriate suppressant, penetrant, and/or surfactant, can be performed.

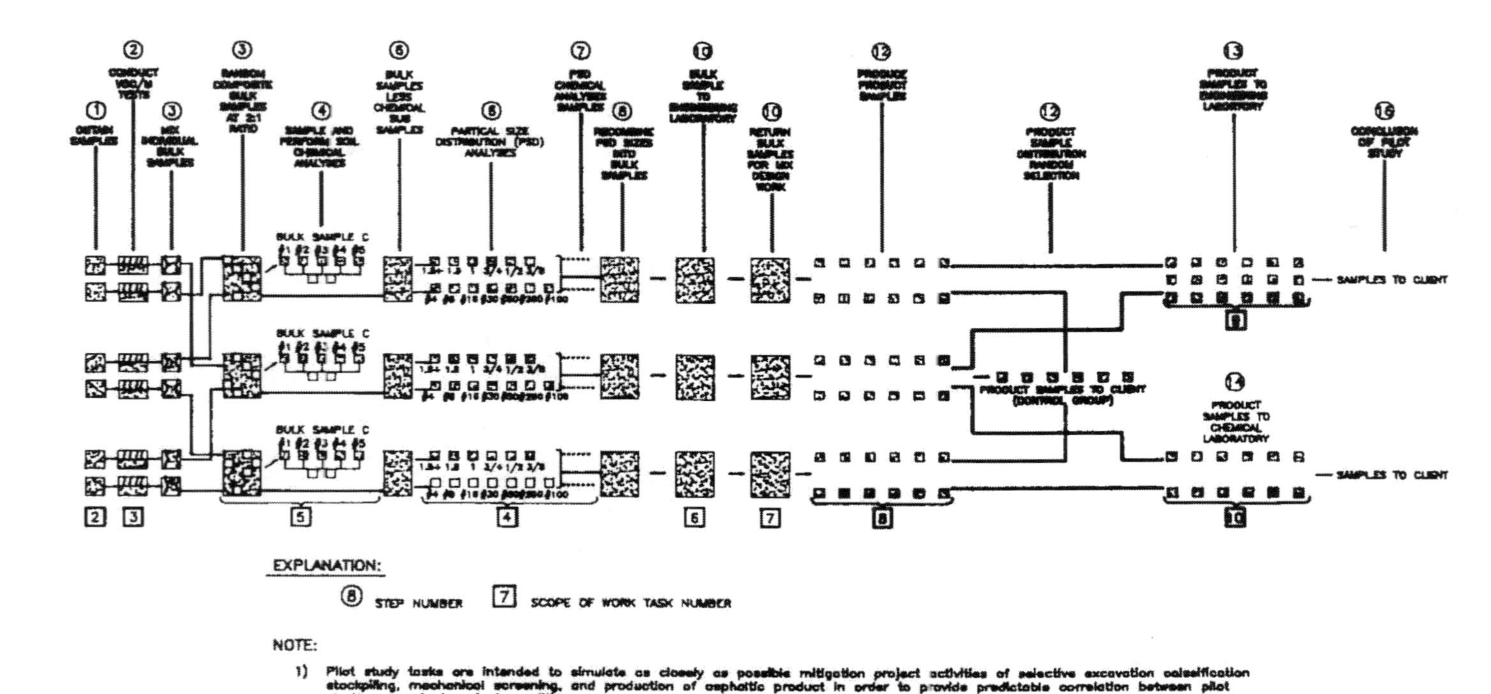

Figure 6-2. Schematic flow chart showing the various sequence of tasks performed during conduct of a pilot test.

PSD is performed to determine whether the contaminants of concern are concentrated within identifiable size fractions that can be segregated in the field, thus reducing the overall quantity of soil being processed. Thirteen gradations result from performing a sieve analysis (1½+, 1½, 1, ½, ⅜, No. 4, No. 8, No. 16, No. 30, No. 50, No. 100, and No. 200 in. sieve sizes). These gradations are then individually tested for the contaminant of concern. A series of chemical tests (Table 5-3) are performed on a worst-case sample to determine not only which constituents are present but also which constituents are not present. Compositing may be implemented. This phase of the pilot test is performed concurrently with PSD analysis. Engineering tests to be performed on the candidate soil that should routinely be incorporated into the pilot program include matrix analysis, density, compaction, R-value, optimum moisture, sand equivalent, expansion index, and any other pertinent tests relevant to site conditions or specific needs of the project. Such tests will be useful in evaluating the options for use of the end product.

An appropriate emulsion will need to be selected. For conventional or routine applications, variation in emulsion grades is minimal. However, it may be prudent to evaluate more than one emulsion grade for purposes of effectiveness and cost. Three to six mix designs should be appropriate for most projects. The more variety in soil types encountered and in end uses will result in the potential need for additional mix designs. Asphalt cores are produced once mix designs are formulated. Once formulated, chemical tests similar to those performed on the aforementioned soil samples are performed. The most important test to be performed on the product samples will be Marshall Stability.

Once all testing is performed, a comprehensive report is typically needed which presents the results of the pilot test program. A cost-benefit analysis should also be included to demonstrate cost-effectiveness of the recycling strategy being pursued. In addition, review of the pertinent regulatory aspects should be presented to assure compliance with federal, state, and local requirements, which undoubtedly will vary from county to county.

The State of New York Department of Environmental Conservation has developed a testing protocol that must be completed for each proposed product and contaminant type prior to issuance of a Beneficial Use Determination (BUD). The data generated is used to set limits of contamination applicable to each producer's specific process and to demonstrate acceptability of the proposed end uses of the material. The protocol also includes material testing to demonstrate the suitability of the finished product for each project performed (i.e., New York State Department of Transportation specifications).

Issues that must be addressed prior to issuance of a BUD include material or contaminated soil testing, type of processing equipment, stockpiling, maximum contaminant levels (MCL) determination, departmental notification, contaminant identification, throughput rates, TCLP testing of processed soil and noncontaminated asphalt analog, and testing requirements for each project to be performed. Strength and durability testing is determined on an as-needed basis. Material testing on the contaminated soil ensures that all materials to be processed will meet the appropriate specifications for grain size, moisture content, and plasticity index, among others. The equipment to be used for application for a BUD must be the same as will be used for full-scale production. The equipment must have the capability to mechanically interlock feed rates for untreated soil and asphalt. All material must also be stockpiled prior to use. When material is placed in layers and is of sufficient size to ensure complete mixing of all the various types of material present, the method for removal of these materials must cut across these layers to ensure proper mixing. Samples are then retrieved from various locations in a face cut into the stockpile (Figure 4-2). MCL testing is performed on the contaminated soil for all contaminant types and ranges that will be present during processing, prior to mixing with aggregate or emulsion. The sample for this test can reflect data previously generated for the project or newly generated data from a recent spill or manufactured contaminated soil specifically composed for a particular project. As with other states, notification to the lead agency of the intent of the project is required. The rate of soil to be processed and air permits, if appropriate, will also be required.

Testing on the processed soil or asphaltic end product for a BUD consists of TCLP testing. The protocol for such testing includes the preparation of four consolidated asphaltic cores. Core numbers 1, 2, and 3 are tested via TCLP after 1, 5, and 10 d, respectively. Three unconsolidated samples are also retrieved from a process pile not less than 5 yd^3 kept covered for 10 d and collected from the pile at no less than 6 in. beneath the surface of the pile. The three samples (sample numbers 1, 2, and 3) are retrieved from the pile after 1, 5, and 10 d, respectively, and subsequently analyzed via TCLP testing. The results obtained are used to evaluate the curing time between mixing of the asphaltic emulsion and material application. Core number 4 is tested for percent asphalt for compliance with specifications. For comparison purposes, a similar testing protocol should also be conducted for the asphaltic product produced using contaminated soil and the same water-based asphalt emulsion formulations.

Upon receiving a BUD, certain testing requirements will have to be met for each project. The producer will be required to test one cured, consolidated product sample in a manner as previously discussed above per project. In addition, such testing must be performed at a minimum rate of one test per 200 tons (or 200 yd^3) of contaminated soil, with no less than two tests performed per project, prior to application and placement on the site of the original contamination only. Curing time must be consistent with that presented prior to receiving the BUD. Should the sample fail such tests even with further processing and/or testing, then the material must be handled as a solid waste.

BIBLIOGRAPHY

Asphalt Institute, 1991, Soils Manual for the Design of Asphalt Pavement Structures, Manual Series No. 10 (MS-10), 4th ed., 259 pp.

Asphalt Institute, 1990, Asphalt Cold Mix Manual, Manual Series No. 14 (MS-14), 3rd ed., 52 pp.

New York State Department of Transportation, 1986, Procedure for Determining Application Rates, Calibration and Inspection for Soil Stabilization Plants, Soil Mechanics Bureau, Soil Control Procedure SCP-10, Official Issuance No. 7.41-3 SCP-10/86, 8 pp.

Smith, R.W., 1987, The Marshall Method for the Design and Control of Asphalt Paving Mixtures, Humboldt Manufacturing Company, Chicago, IL, 62 pp.

7 CHEMICAL ASPECTS OF BITUMENS AND ASPHALT

7.1 INTRODUCTION

The use of HMA or CMA asphalt to stabilize contaminants in organic- and inorganic-affected soil using soil reuse and recycling technologies has been shown to be a viable and creative method of utilizing affected soil to produce a useful end product instead of a waste requiring disposal. Evaluation of asphalt has traditionally focused on structural performance as pavement and building materials. With the incorporation of affected soil as part of the aggregate, various studies conducted over the past 3 decades have shown these materials not to adversely affect the structural behavior of the asphalt as long as general requirements are met (i.e., the mix design must meet its end use requirements). However, evaluating the long-term performance of an asphaltic product incorporating affected soil, notably when the end product is used as a liner, designed fill, or other surface or subsurface use, must focus on chemical performance and requires more extensive experimental study.

Petroleum products and asphalt components are generally immiscible with respect to the aqueous phases expected under subsurface conditions, with the exception of polar functional groups. These groups or components assist in preventing the release of contaminants from the asphalt and slowing diffusion of aqueous components out of the asphalt. Therefore, the release of any incorporated contaminant (i.e., petroleum, polynuclear aromatics, metals, etc.) from the asphalt must be coupled with the hydraulic conductivity, diffusivity, and structural breakdown of the asphalt itself, which are for all practical purposes very slow processes. For similar crude sources and chemical treatments, the end product asphalt cement formed from either cold-mix or hot-mix emulsion has the same general structure, composition, and properties. The greatest effects between the different preparations is on kinetically controlled reactions and sorption processes.

Presented in this chapter is discussion of types of bitumens, asphalt-aggregate chemistry, including important functional groups, and asphalt-aggregate stability. Contaminant mobility, permeability, and leachability from asphalt, in addition to other important factors to be considered, such as durability, aging, and biological resistance of the asphaltic end product, are also discussed.

7.2 TYPES OF BITUMENS

Bitumens have been used for thousands of years. Bitumens have been known to be used as a cementing agent for various building materials in Babylon and has also been used for waterproofing and the caulking of boats. Its probable earliest use was as a water stop between brick walls of a reservoir in the beginning of the third millennium B.C. at Mohenjo-Daro, India. Reference is also made to the use of asphalt as mortar in Genesis XI, 3. Early Buddhists refer to earth-butter; in the Middle East asphalt was extensively used for roads and water works such as flood control. It is stated that a king left an inscription that stated he had "found his realm in mud and left it laced with roads glistening in asphalt."

The deposits of bitumens in natural asphalts are known to occur in many parts of the world, including California, Canada, Cuba, Venezuela, and Syria. These deposits have been designated as bitumen, asphaltite, pitch, elaterite, albertite, gilsonite, impsonite, grahamite, and wurtzilite. Deposits of glance pitch in Utah have been reported to contain an average of 1.75% uranium oxide (U_3O_8) and 4% vanadium pentoxide (V_2O_5).

Bitumens and bituminous materials are currently used for a variety of purposes. In the United States, about 75% of all bitumens are used for paving, with about 15% used in roofing and the remainder for a variety of other purposes (i.e., lining of canals and reservoirs, injection into sands and fractured rock to serve as a barrier to groundwater flow, insulation, coatings, paints, inks, etc.).

Bituminous substances are a class of native and pyrogenous substances containing or treated with bitumens or pyrobitumens, or substances of similar physical properties. Included in this broad group of substances are bitumens, pyrobitumens, pyrogenous distillates (i.e., pyrogenous waxes and tars), and pyrogenous residues (i.e., pitches and pyrogenous asphalts including asphalt and asphaltic pyrobitumens). Community products include bituminous concrete, bituminized felts and fabrics, and bituminous pavement. The relationship between the various groups of bituminous substances is determined by such factors as origin, volatility, composition, mineral matters, fusibility, and solubility.

7.2.1 Physical and Chemical Criteria

In the United States, the common term "asphalt" is a species of "bitumen," a term broader in scope and more commonly used in Europe. Bitumens are defined as mixtures of hydrocarbons that naturally occur in nature or are obtained as residual material through distillation in the refinery process. The Egyptians used bitumens, derived from the Latin term "pix tumens," meaning "tomb's pitch," from the Black Sea for the embalming of corpses. The word bitumen comes from the original Sanskrit word Gwitu-men, meaning pitch. More specifically, ASTM defines bitumen as mixtures of hydrocarbons of natural or pyrogenous origin, or combinations of both, frequently accompanied by their nonmetallic derivatives, which may be gaseous, liquid, semisolid or solid, and which are completely soluble in carbon disulfide. Bitumens are composed principally of high molecular weight hydrocarbons, such as asphalts, tars, pitches, and asphaltites.

Pyrobitumen is a generic term for a dark colored, comparatively hard and nonvolatile, native substance composed of hydrocarbons. Such substances may or may not contain oxygenated bodies, sometimes associated with mineral water. The nonmineral constituents are infusible and relatively insoluble in carbon disulfide. This definition also applies to asphaltic and nonasphaltic pyrobitumens, and their respective shales.

Bitumen is organic material consisting of quite variable mixtures of mainly aliphatic and aromatic hydrocarbons, combined in part with nitrogen, sulfur, and oxygen, of high molecular weight. Analysis usually consists of determining molecular weights, general nature of compounds, and elemental composition, rather than specific chemical structure. Bitumens derived from natural deposits resemble those derived from the distillation of petroleum, although natural bitumens often contain minerals, whereas those derived from refineries generally have more paraffinic side-chains than natural bitumens. Crude oils can contain up to 50% bitumens.

Mineral wax refers to a species of bitumen and to certain pyrogenous substances. Mineral wax is characterized by variable color, viscous to solid consistency, characteristic luster and unctuous feel, and comparative nonvolatility. Composed of hydrocarbons, mineral wax is virtually free from oxygenated bodies and contains considerable crystallizable paraffins, some associated with mineral matter. The nonmineral constituents are easily fusible and soluble in carbon disulfide. This definition also applies to crude and refined native mineral waxes and pyrogenous waxes.

Asphalt is a species of bitumen. Asphalt is a generic term for a dark brown to black cementitious material, solid or semisolid in consistency, in which the predominant constituents are bitumens (i.e., high molecular weight hydrocarbons) that occur in nature as such or are obtained as residual materials in the refinery process. The Greek work for native asphalt is "asphaltos." Asphalts are characterized by variable hardeners, comparatively nonvolatile substances composed of hydrocarbons that are substantially free of oxygenated bodies. Asphalts contain relatively little to no crystalline paraffins. Sometimes associated with mineral matter, nonmineral constituents are fusible and largely soluble in carbon disulfide. Distillate, which fluctuates between 300 and 350°C, yields considerable sulfonation residue. This definition also applies to native asphalts and pyrogenous asphalts.

Asphaltite is a species of bitumen including dark colored, comparatively hard and nonvolatile solids. Asphaltite differs from asphalt in that the nonmineral constituents are not easily feasible although largely soluble in carbon. This definition applies to gilsonite, glance pitch, and grahamite.

Asphaltic pyrobitumens are a species of pyrobitumen, are characterized by dark colored, comparatively hard and nonvolatile solids composed of hydrocarbons. Substantially free of oxygenated bodies, asphaltic pyrobitumens are sometimes associated with mineral matter. Nonmineral constituents are refusible and largely insoluble in carbon disulfide. This definition applies to elaterite, wurtzilite, albertite, impsonite, and asphaltic pyrobituminous shales.

7.2.2 Mode of Preparation Criteria

Bitumens can also be divided according to their mode of preparation. Bitumen types by mode of preparation include

- straight-run distillation bitumens
- oxidized bitumens
- cracked bitumens
- cutback bitumens or asphalt
- emulsified bitumens
- natural bitumens

Straight-run distillation bitumens are obtained as residue after distillation of certain petroleum crude oils. Oxidized bitumens are obtained by blowing air through molten bitumens at 150 to 260°C. Cracked bitumens are generated by breakdown of high molecular weight compounds including cracked residues. Cutback bitumens or asphalt are generated by liquefying bitumens by the addition of a solvent (i.e., petroleum distillates). Emulsified bitumens are generated by the addition of anionic (alkaline soap); cationic (amine salt); or nonionic aqueous solutions (emulsifiers). Natural bitumens consist of naturally occurring materials and include asphalts, asphaltic pyrobitumens, rock asphalt, and glance pitch.

7.3 ASPHALT–AGGREGATE CHEMISTRY

Asphalt is a dark, brown to black cementitious material, solid or semisolid in consistency, in which the predominating constituents are bitumens (i.e., high molecular weight hydrocarbons) that occur in nature as such or are obtained as residuals in the refining process. Asphalt has a complex and poorly understood chemistry and structure. Its chemistry and structure largely depends upon the crude petroleum source and any chemical treatment and/or chemical modifiers added during processing. Asphalt has a large number of heteroatomic groups with a wide range of chemical reactives. Therefore, it seems likely that a number of asphalt, hydrocarbon, or metal–soil reactions that could potentially affect contaminant mobility might occur in contaminated soil subjected to reuse and recycling methodologies.

7.3.1 Functional Groups

Important functional groups found in asphalts include polynuclear aromatics, phenolics, 2-quinolines, pyrrolics, pyridinics, sulfides, sulfoxides, anhydrides, carboxylic acids, and ketones as shown in Figure 7-1. The number and distribution of these groups vary widely among different asphalts and determine much of their chemical behavior. Of these groups, carboxylic acids, ketones, and anhydrides are generally formed by atmospheric oxidation (i.e., ketones and anhydrides) or caustic pretreatment (i.e., carboxylic acids concentrated in treated asphaltic residues) and are rare in fresh asphalts. However, these may be important in waste disposal situations because of possible interactions with oxidizing or alkaline contaminant solutions, or if recycled asphalt is used.

7.3.2 Functional Group Analysis

There has been no comprehensive study of asphalt chemistry in relation to the aggregate or to contaminant species; therefore, generalizations about the chemical performance of these systems is difficult. However, extrapolations from asphalt studies of road pavement properties, leaching behavior, sensitivities to moisture-damage, and functional group analysis have provided information that can be used to evaluate the stability of hydrocarbon compounds, metals, and other contaminants in soils that have been incorporated into asphalt.

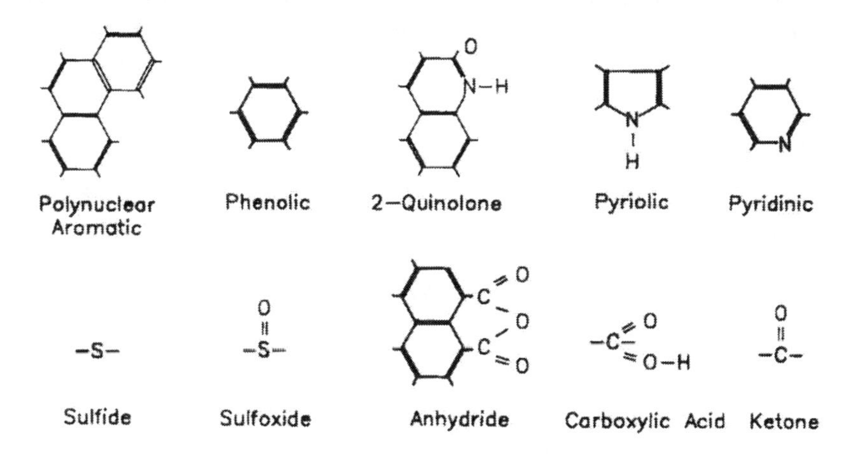

Figure 7-1. Important polar and nonpolar functional groups present in asphalt. (Modified from Conca and Testa, 1991 and Testa and Conca, 1993.)

It is generally assumed that asphalt cements are colloidal systems made up of a suspension of asphaltene micelles in an oily medium. Micelles are units of various molecules, usually organic with minor inorganic components, that have distinct structural and/or chemical properties. Micelles play an important role in asphalt behavior. The structural and chemical components in affected soil that has been asphalted are illustrated in Figure 7-2. The figure is prepared approximately to scale and represents the results and implications of separate studies of asphalt components. Major components include aggregate mineral grains (i.e., clay mineral), bulk aqueous phase, sorbed electric double layer on the mineral surface, resin-peptized asphaltene micelles, polar micelles in the asphalt, asphalt oily medium, asphalt functional groups at the interfaces between these different components, and asphalt pore spaces. Each component will be described separately in reference to Figure 7-2.

The aggregate mineral grain portrayed in Figures 7-2 and 7-3 is a clay mineral. Clay minerals are important because of their reactive surfaces and their ability to exchange cations from the interlayer sites with contaminant ions in solution. There are two major types of complexing functional groups associated with silicate mineral grains. Complexation (or chelation) is the process by which metal ions and organic or other nonmetallic molecules, referred to as ligands, can combine, forming stable metal–ligand complexes. The most important is the siloxane ditrigonal cavity, which occurs in tetrahedral silicate sheets and gives the clay minerals their exchange capacities. However, the most abundant surface functional group is the inorganic hydroxyl group, which is found on almost all mineral and amorphous solid phases. Hydroxyl groups exposed on tetrahedral silica are called silanols, whereas those exposed on octahedral alumina are called aluminols. There can be more than one type of surface hydroxyl on a given mineral surface with different reactives. For example, the siloxane cavity complexes only positive ions and groups, whereas the surface hydroxyl complexes both anion and cation species, depending on solution composition and pH. Asphalt functional groups as well as contaminant species will interact with these surface complexes. Several mineral surface interactions are illustrated in Figures 7-2 and 7-3, including a strong complexation with a uranyl anion complex (Figure 7-4), a weak complexation with a hydrated calcium complex, an exchange of americium with an interlayer cation, sorption of some short-chain hydrocarbons at the mineral surface, and bonding of the asphalt quinoline, pyrrolic, phenolic, and carboxylic acid functional groups.

The bulk aqueous phase has many of the dissolved inorganic constituents of interest to contaminant transport. Because the simple diffusion coefficient of dissolved species in the bulk fluid is approximately 10^{-5} cm^2/s regardless of species, diffusion of contaminants through the bulk aqueous phase should be the primary route of contaminant release from the asphalt. The release rate, however, depends strongly on the connectivity of the bulk fluid (i.e., the diffusion porosity). In coherent asphalts without moisture damage, this

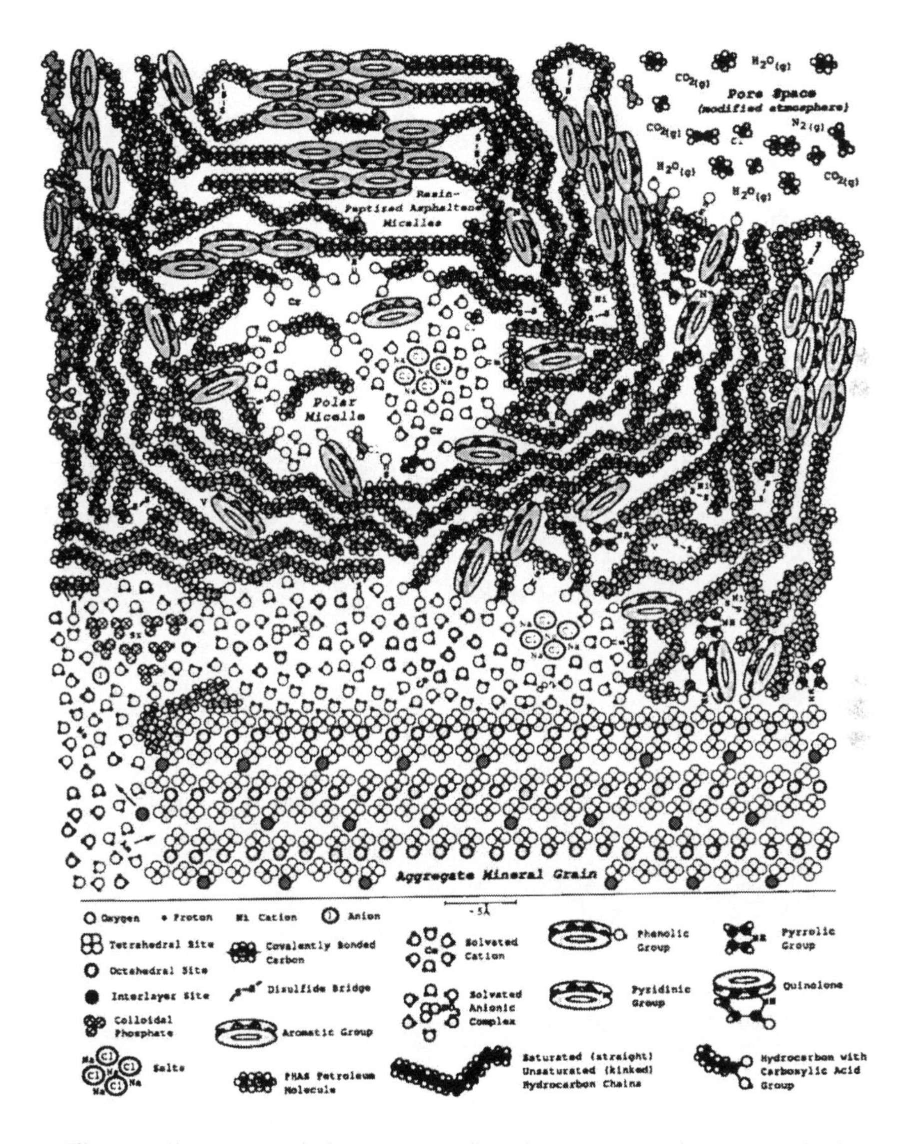

Figure 7-2. The asphalt–aggregate interface at the molecular level using space-filling and structural representations to illustrate important functional relationships. (Modified from Conca and Testa, 1992 and Testa and Conca, 1993.)

connectivity is small, and the effective diffusion coefficient in the asphalt cement as a whole is low (on the order of 10^{-12} cm^2/s). Release of contaminant species also depends strongly on the retardation properties of the system. Other phases in contact with the bulk aqueous phase can sorb species that

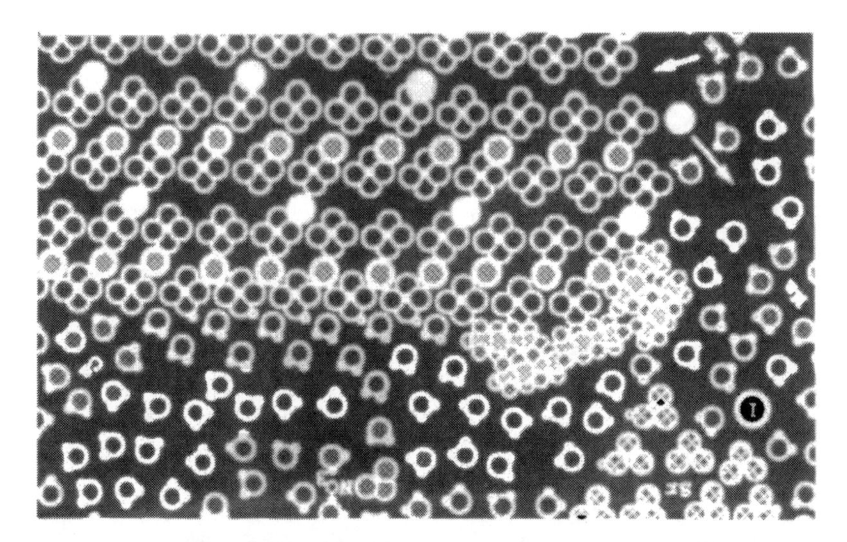

Figure 7-3. Space-filling and structural relationship illustrating various anticipated surface interactions of a clay mineral incorporated as aggregate into asphalt.

Figure 7-4. Space-filling and structural relationship illustrating interaction between a clay mineral and uranyl anion complex (UO_2) and certain functional groups.

are present in the aqueous phase. The strong sorption of strontium and the weak sorption of iodine to a phosphate colloid particle in the bulk aqueous phase are also illustrated in Figure 7-3.

The asphaltene micelles and the oily medium make up the bulk of the asphalt. Asphaltenes are molecules composed of polynuclear aromatic groups and long hydrocarbon chains which contain much of the inorganic constituents of asphalt. An asphaltene micelle is an aggregate of asphaltene molecules bonded through μ-electron and interactions between the condensed polynuclear aromatic sheets. Asphaltene micelles can be peptized by hydrocarbon resins into aggregates, as illustrated in Figure 7-2. The oily medium consists of saturated and unsaturated hydrocarbons and resins. These resins are a mixture of terpenes, resin alcohols, and resin acids and their esters, the complexity of which are not represented in Figure 7-2. There may be extensive secondary structures to the micelles that could have great structural importance. Certain metals, such as vanadium, nickel, and iron, which are found in asphalts, are thought to be associated with the asphaltenes, although exact molecular sites are not known. They may, however, be associated with sulfides and aromatic groups and are represented as such in Figure 7-5.

The possible polar compounds present in polar micelles are also illustrated in Figure 7-2. Evidence has been documented showing that agglomerates or micelles of polar asphalt molecules exist separately in the asphalt and sequester certain polar complexes and acids, such as manganese acid complexes. Carboxylic acid groups will certainly occur in the polar micelles if they exist in a particular asphalt, and polar functional groups such as the phenols, ketones, and sulfoxides will be concentrated at the interface. This

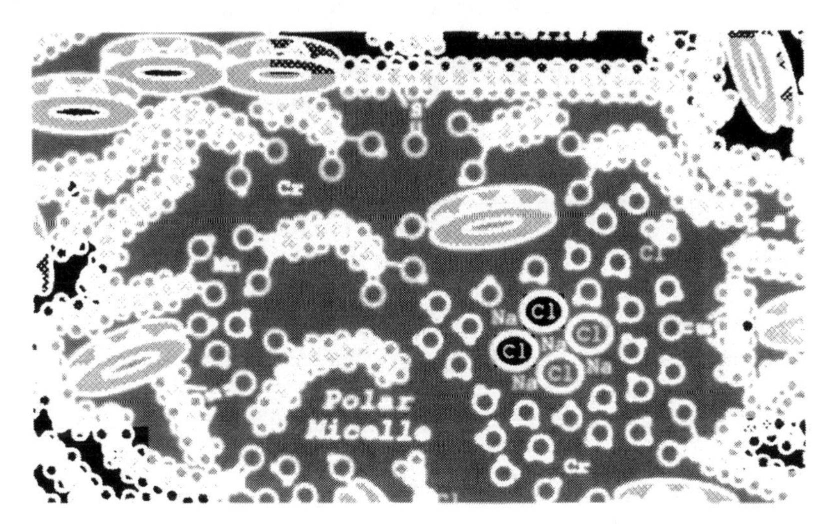

Figure 7-5. Space-filling and structural relationship illustrating interaction between certain metals and aromatic groups.

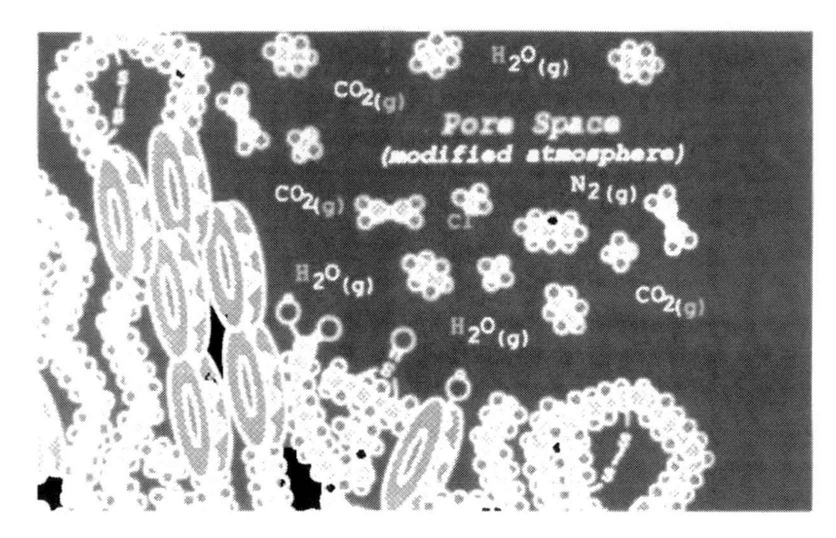

Figure 7-6. Space-filling and structural relationship illustrating interaction showing sequestering of water molecules.

sequestering chemically removes those species from subsequent reactions in the nonpolar phases and aqueous phases of the asphalt. The polar micelles could sequester many metals species of interest. Speculation exists that water molecules and salts can occur in the polar micelles; this has been incorporated into Figures 7-5 and 7-6.

The nature of the pore space in asphalt is a highly debated topic. Although pore space is limited with no evident connectivity, as is shown in a petrographic thin section (Figure 4-1), there is a great deal of pore space in asphalt that can be observed in electron micrographs, but again these are not well connected. Depending upon the conditions of formation and subsequent history, the pore space may or may not be water filled. The gaseous components in pore spaces will include atmospheric components introduced during formation and processing and volatile asphalt compounds. Gaseous contaminants of interest may also be incorporated into the pore spaces.

Knowing the structures and components of asphalt incorporating affected soil is crucial to understanding its chemical performance with respect to hydrocarbon and metal contaminants. This is exemplified by several examples. Heavy metal salts of carboxylic acids ($RCOO^-M^+$), which are insoluble in water, are soluble in strong acids and in strong polar solvents, making polar micelles possible repositories of contaminant metal–carboxylic acid salts in the asphalt incorporating affected soil. Petroleum contaminants in the affected soil will also be strongly hydrogen-bonded to the asphalt components as well as possibly bonded through reactions with the functional groups. Furthermore, oxidation-reduction reactions of metals will also be affected by the polar and non-polar phases in the asphalt incorporating affected soil. Chromium in immiscible cresol/water mixtures has been found to preferentially enter the non-aqueous cresol phase (isomeric phenols) where the oxidation-reduction reaction Cr(VI)→Cr(III) occurred much faster than in the aqueous phase. This effect has important implications for metals in affected soil, especially chromium interactions with asphaltic phenolic groups. However, these reactions are kinetically controlled and may be unimportant at pH 8.

7.4 ASPHALT–AGGREGATE STABILITY

Incorporation of affected soil as an ingredient to produce an asphaltic product is achieved by stabilization, solidification, and encapsulation. Stabilization is where chemical fixation techniques render a waste less toxic or harmful to the environment. The hazard potential of what was once considered a waste is subsequently reduced. Examples of such techniques are ion exchange of heavy metals in an alumina silicate matrix of a cementitious agent or sorption of heavy metals on fly ash in an aqueous system, among others. Stabilization of asphalt incorporating affected soil can be functionally described in terms of the final product, in which the constituents of concern are mechanically stabilized and immobilized by fixation and isolation. This differs from solidification, where the waste is transformed into a stable and durable matrix that is more compatible for reuse, storage, or disposal. Solidification creates barriers between the waste components and the environment by either reducing the permeability of the waste and/or matrix, or reducing the effective surface area available for diffusion, or both, with or without chemical fixation.

The chemical stability and environmental performance of asphalt incorporating affected soil are, however, dependent upon the nature and extent of the asphalt–aggregate bonds. Much of the information concerning these bonds comes from studies of asphalt sensitivity to moisture-induced damage. Because of the polar/nonpolar and hydrophobic/hydrophilic interactions among the various phases in the system, the functional groups, shown in Figure 7-1, will be concentrated at the interfaces between phases (i.e., asphalt–aggregate interface, polar micelle–asphalt oil interface, asphalt–water interface, etc.). Stability is thus addressed via the functional groups' relative tendency to be concentrated at the asphalt–aggregate interface, sorption affinity, and ability to be displaced by water.

The behavior of functional groups at the asphalt–aggregate interface determines the chemical stability of the asphalt to a large degree. At the asphalt–aggregate interface, these groups are susceptible to interactions with the aqueous phases and any contaminants present in the affected soil. The relative tendency of the asphalt functional groups to be concentrated at the asphalt–aggregate interface has been determined, in order of decreasing tendency, as

carboxylic acid > anhydride >> quinoline, phenolic > sulfoxide, ketone > pyrrolic

The sorption affinity of the functional groups with the aggregate surface has also been investigated and is given, in order of decreasing tendency, as

pyridinic, carboxylic acid >> anhydride > quinoline,
phenolic > sulfoxide > ketone >> pyrrolic, polynuclear aromatic

The third important tendency of the functional groups affecting asphalt incorporation stability is the ability of the functional group-aggregate bond to be displaced by water. This tendency, listed in order of decreasing tendency, is given as

anhydride, quinoline, carboxylic acid > sulfoxide >
ketone > pyrrolic, phenolic

Combining these tendencies provides some indications of performance and some guidelines for using asphalt incorporation methodology. For the best chemical performance, the asphalt should have high contents of pyridinic, phenolic, and ketone groups, which can be achieved by selectively choosing the source material or using additives such as shale oil. Also, the presence of inorganic sulfur, monovalent salts, and high ionic strength solutions in the asphalt decreases the chemical stability of the asphalt cement by disruption of the functional group-aggregate bonds, and increases the system's permeability. Addition of lime to the aggregate can thus be used to counter this effect.

7.5 CONTAMINANT MOBILITY IN ASPHALT

Contaminant mobility, especially of metals, in the asphalted cement will be affected by many factors, including diffusivity, permeability, solubility, specification, complexation, redox reactions, sorption, and precipitation. These factors' influence on contaminant mobility in asphalt are summarized below:

- Diffusivity and permeability of the asphalt cement (i.e., as a whole generally less than 10^{-12} cm^2/s and 10^{-9} cm/s, respectively;
- Solubility of species in the various aqueous, polar, and nonpolar phases;
- Specification of the contaminants (i.e., UO^2 [nonmobile] vs. $UO^2 \cdot nCO^3$ [highly mobile]);
- Complexation with any chelating organics;
- Redox reactions of metals across aqueous–organic phase boundaries;
- Sorption on aggregate surfaces, along asphalt–aqueous interface, or on colloids; and
- Precipitation of solid phases and/or colloids of metal salts, especially oxyhydroxide and carboxylic acid salts.

Many of these properties are not known to the degree that specific contaminant release can be predicted. Leaching tests, the primary method of evaluating contaminant mobility in these systems, have been performed on a variety of asphalts. However, the low diffusivities and permeabilities of asphalt are obviously the greatest factor in the retention of contaminants in asphalt cements. Conditions that adversely affect the diffusivity and permeability will have the greatest adverse effect on contaminant mobility and release. The asphalt acts primarily as a physical containment (i.e., via fixation or stabilization, solidification, and encapsulation) to the contaminants and the aggregate, depending upon the asphalt cement's overall composition and structure.

7.6 CONTAMINANT LEACHABILITY FROM ASPHALT

Even though metals such as vanadium and nickel occur in asphalts in the hundreds of parts per million levels, as well as many toxic organic components, asphalt leachates and products have never produced toxic or contaminated solutions that are considered to be hazardous by the U.S. EPA. Asphalt's nonvolatile, viscoelastic properties result in the general observation that asphalt leachates do not contain reported contaminant concentrations above the U.S. EPA's drinking water guidelines. Even in asphalts incorporating metal slags as an aggregate, metals do not become solubilized and do not leach from these asphalts in detectable concentrations when used with strong acidic and alkaline solutions. Some typical test results to assess the potential leachability of certain hydrocarbon compounds- and metals-affected soil incorporated as an ingredient in HMA and CMA are presented in Tables 7-1 and 7-2, respectively.

During evaporation phases and compaction, void structure collapses and water leaves the system. Although leachability of contaminants is an issue to be addressed, leachability studies indicate that contaminants are nonleachable during these various phases.

Asphalt leachates that have contained detectable concentrations of contaminants have been obtained in studies of asphalted nuclear wastes. However, in all of these studies, high concentrations of salts ranging up to 50% salt to 50% asphalt were used. It is known that high salt concentrations in asphalt mixes disrupt the asphalt structure, a condition that will not occur in affected soil subjected to asphalt incorporation methodologies.

On the other hand, in diffusion studies of radioactive wastes with normal salt contents, diffusion coefficients were measured to be as low as for normal asphalt conditions; with diffusion coefficients of 10^{-12} cm^2/s and 10^{-13} to 10^{-10} cm^2/s. In all these studies, researchers point out that experimental effects (i.e., slicing of thin asphalt membranes) may introduce errors that are not relevant to the field situation and tend to increase the observed diffusion coefficient.

A number of unplanned leaching experiments have been taking place with HMA and CMA asphalt. Asphalt has been used for years to line domestic

Table 7-1 Summary of Leachability Test Results from Seven Reclaimed Pavement HMA Using U.S. EPA Test Procedures[a,b]

Parameter	Sample No.						
	1	2	3	4	5	6	7
Barium[c]	<0.2	0.40	0.36	0.33	<0.2	<0.2	<2
Cadmium[c]	<0.2	<0.2	<0.2	<0.2	<0.2	<0.2	<0.02
Chromium (III)[c]	<0.05	0.52	<0.05	<0.05	<0.05	<0.05	0.10
Lead[c]	<0.2	1.80	<0.2	<0.2	<0.2	<0.2	<0.2
Silver[c]	<0.04	<0.04	<0.04	<0.04	<0.04	<0.04	<0.04
Arsenic[c]	<0.005	<0.005	<0.005	<0.005	<0.005	<0.005	<0.005
Selenium[c]	<0.025	<0.025	<0.025	<0.025	<0.025	<0.025	<0.025
Mercury[c]	<0.005	<0.005	<0.005	<0.005	<0.005	<0.005	<0.005
1,4 Dichlorobenzene[d]	<50	<50	<50	<50	<50	<50	<12
2,4 Dinitrotoluene[d]	<50	<50	<50	<50	<50	<50	<12
Hexachlorobenzene[d]	<50	<50	<50	<50	<50	<50	<12
Hexachlorobutadiene[d]	<50	<50	<50	<50	<50	<50	<12
Hexachloroethane[d]	<50	<50	<50	<50	<50	<50	<12
Nitrobenzene[d]	<250	<250	<250	<250	<250	<250	<12
Pyridine[d]	<120	<120	<120	<120	<120	<120	<60
Cresylic acid[d]	<50	<50	<50	<50	<50	<50	<30
2-Methyl phenol[d]	<50	<50	<50	<50	<50	<50	<30
3-Methyl phenol[d]	<50	<50	<50	<50	<50	<50	<30
4-Methyl phenol[d]	<250	<250	<250	<250	<250	<250	<30
Pentachlorophenol[d]	<250	<250	<250	<250	<250	<250	<60

2,4,5-Trichlorophenol[d]	<50	<50	<50	<50	<50	<50	<30
2,4,6-Trichlorophenol[d]	<50	<50	<50	<50	<50	<50	<30
Naphthalene[d]	0.49	<0.13	<0.13	0.30	<0.13	<0.13	0.25
Acenaphthylene[d]	<0.20	<0.20	<0.20	<0.20	<0.20	<0.20	<0.15
Acenaphthene[d]	0.14	<0.13	<0.13	<0.13	<0.13	<0.13	<0.194
Fluorene[d]	<0.015	<0.015	<0.015	<0.015	<0.015	<0.015	<0.023
Phenanthrene[d]	<0.13	<0.13	<0.13	<0.13	<0.13	<0.13	<0.023
Anthracene[d]	<0.017	<0.017	<0.017	<0.017	<0.017	<0.017	<0.015
Fluoranthene[d]	<0.017	<0.017	<0.017	<0.017	<0.017	<0.017	<0.037
Pyrene[d]	<0.060	<0.060	<0.060	<0.060	<0.060	<0.060	<0.04
Benzo(A)Anthracene[d]	<0.017	<0.017	<0.017	<0.017	<0.017	<0.017	<0.048
Chrysene[d]	<0.033	<0.033	<0.033	<0.033	<0.033	<0.033	<0.017
Benzo(B)Fluoranthene[d]	<0.023	<0.023	<0.023	<0.023	<0.023	<0.023	<0.02
Benzo(K)Fluoranthene[d]	<0.017	<0.017	<0.017	<0.017	0.050	<0.017	<0.022
Benzo(A)Pyrene[d]	<0.024	<0.024	<0.024	<0.024	<0.024	<0.024	<0.023
Dibenzo(A,H)Anthracene[d]	<0.068	<0.068	<0.068	<0.068	<0.068	<0.068	<0.018
Benzo(G,H,I)Perylene[d]	<0.110	<0.110	<0.110	<0.110	<0.110	<0.110	<0.036
Indeno(1,2,3-CD)Pyrene[d]	<0.022	<0.022	<0.022	<0.022	<0.022	<0.022	<0.021

[a] Analytical tests include U.S. EPA Methods SW846-3350, 8080, 1311, 3510, 8310, and 3010.

[b] Symbol < denotes below given analytical detection limit.

[c] ppm = parts per million.

[d] ppb = parts per billion.

From Kriech, A. J., 1990, Evaluation of Hot Mix Asphalt for Leachability, Heritage Research Group HRG 3959A0M3, Indianapolis, IN.

Table 7-2 Summary of Leachability Test Results for Bulk Preprocessed Soil Samples and Processed Cold-Mix Asphalt

Parameter (EPA method)	Preprocessed soil sample number				Maximum contaminant level[b]	Leach test results on processed asphalt
	1	2	3	4		
Total recoverable petroleum hydrocarbon (418.1)	375	5120	775		No regulatory levels established	ND[c]
Total petroleum hydrocarbon (8015M)	730	1350	79	13,000	No regulatory levels established	ND
Total petroleum hydrocarbon 8015M leachate	ND	ND	ND		No regulatory levels established	ND
Semivolatile organic compounds (8270) detectable constituents						
Acenaphthene	7	4.9	ND		No TCLP[d] regulatory levels have been established for constituents detected by these laboratory analyses	ND
Acenaphthylene	78	17.4	11			ND
Anthracene	56	11.6	6			ND
Benzo[a] Anthracene	54	15.7	5			ND
Benzo[b] Fluoranthene	79	22.2	ND			ND
Benzo[k] Fluoranthene	45	11.0	ND			ND
Benzo[a] Pyrene	63	20.3	ND			ND
Benzo[g,h,i] Perylene	31	15.5	ND			ND
Chrysene	95	19.7	11			ND
Fluoranthene	150	38.5	47			ND

Fluorene	45	10.8	3					ND
Indeno[1,2,3-cd] Pyrene	20	10.4	ND					ND
2-Methylnapthalene	5	0.8	ND					ND
Napthalene	105	14.1	8					ND
Phenanthrene	101	23.4	20					ND
Pyrene	163	39.2	57					ND
					CCR Title 22 regulatory levels			
CAM Title 22 metals (TTLC)[d]					TTLC[e]	STLC[f]	TCLP	
Antimony	ND	ND	ND	720	500	15	NLE[g]	ND
Arsenic	7.9	5.9	2.4	7000	500	5.0	5.0	ND
Barium	58.9	50.3	66.2		10,000	100	100	ND
Beryllium	ND	ND	ND		75	0.75	NLE	ND
Cadmium	3.9	4.6	1.75		100	1.0	1.0	ND
Chromium (total)	13.7	13.3	15.2		2500	5	5.0	ND
Cobalt	7.9	4.75	4.40		8000	80	NLE	ND
Copper	37.2	26.6	12.4	27,000	2500	25	NLE	ND
Lead	35.4	80.0	5.35	450	1000	5	5.0	ND
Mercury	1.90	1.0	1.80		20	0.2	0.2	ND
Molybdenum	ND	ND	ND		2	3500	350	ND
Nickel	76.6	11.2	7.95		2000	20	NLE	ND
Selenium	ND	ND	ND		100	1	1.0	ND
Silver	ND	ND	ND		500	5	5.0	ND
Thallium	ND	ND	ND		700	7	NLE	ND
Vanadium	40.5	24.0	22.0		2400	24	NLE	ND
Zinc	43.3	63.4	31.6		5000	250	NLE	ND

Table 7-2 Summary of Leachability Test Results for Bulk Preprocessed Soil Samples and Processed Cold-Mix Asphalt *(continued)*

Parameter (EPA method)	Preprocessed soil sample number				Maximum contaminant level[b]			Leach test results on processed asphalt
	1	2	3	4	CCR Title 22 regulatory levels			
					TTLC[e]	STLC[f]	TCLP	
CAM Title 22 metals (STLC)[h]								
Lead[i]		1.86			1000	5		ND
Flashpoint: Ignitability (1010)	>220°F	>220°F	>220°F		>140°F Ref: CCR Title 22 Sec. 66261.21			
pH: Corrosivity (9045)	4.06	3.75	3.52		Less than 2 units or greater than 12.5 units Ref: CCR Title 22 Sec. 66261.22			
Cyanide: Reactivity (9010)	32	51	26		250 mg/kg Ref: CalEPA DTSC, Mr. Ron Piloran, personal communication, 11/4/92			
Sulfide: Reactivity (376.2M)	60	72	56		500 mg/kg Ref: CalEPA DSTC, Mr. Ron Piloran, personal communication, 11/4/92			

Acute Aquatic 96-h LC50 Bio Assay: Toxicity[h]	100% Passing	90% Passing	90% Passing	Ref: CCR Title 22 Sec 66261.24[a,b]

[a] EPA = United States Environmental Protection Agency.

[b] Concentrations expressed as mg/kg or equivalent to parts per million unless otherwise noted.

[c] ND = Not detected above its respective analytical detection limit.

[d] TCLP = Toxicity Characteristic Leaching Procedure (CCR Title 22 Section 66261.24).

[e] TTLC = Total Threshold Limit Concentration (CCR Title 22 Section 66261.24).

[f] STLC = Soluble Threshold Limit Concentration (CCR Title 22 Section 66261.24).

[g] NLE = No level established. TCLP levels for this element not established by Regulatory Agency.

[h] CAM Title 22 Metals Laboratory Testing Protocol per Cal EPA Waste Classification Unit Guidelines. Any element indicating a TTLC concentration of ten times the STLC Regulatory Level should be analyzed by STLC Methods. The only sample requiring STLC testing was 2, for lead. No other elements were in excess of ten times STLC levels expressed as TTLC concentrations.

[i] CCR Title 22, Sec. 66261.24[a,b] requires wastes to pass the 96-h Aquatic Toxicity testing with greater than a 50% survival rate at 500 mg/l concentration for compliance with nonhazardous criteria. Samples 1, 2, and 3 passed with a less than 10% mortality at waste concentration of 750 mg/l, which is a higher waste concentration by 50% than that required. By CCR Title 22 definition "Acute Aquatic 96-Hour LC50" means the concentration of a substance or mixture of substances in water, in milligrams per liter, which produces death within 96-h in half of a group of at least 10 fish.

water reservoirs, especially in California, and to line fish-rearing ponds, with no observable adverse effects. There are over 30 asphalt-lined fish-rearing ponds throughout the states of Oregon and Washington. Trace metal and organic contamination are highly toxic to fry and developing fish. Yet, no adverse effects have been observed from the asphalt liners, indicating a high degree of chemical stability with respect to aqueous solutions and an absence of any toxicity effects.

The physical properties of asphalt depend on functional groups, which vary significantly. The thermodynamics of these functional groups is basically unknown for asphalt. In addition, diffusivity coefficients and conductivity are so low for asphalts that the issue is basically irrelevant. An attempt is made, however, to address those physical properties as they relate to certain engineering considerations.

The testing and performance assessment of asphalt have traditionally focused on its structural performance as pavement and building material. However, when evaluating the long-term performance of asphalt liners produced with affected soil as part of the aggregate, the focus is on their chemical performance. Because of the great immiscibility of petroleum products with respect to the aqueous phases expected under impoundment conditions, large favorable free-energy exchange exists for preventing the release of contaminants from the asphalt; therefore, the chemical behavior and performance of the petroleum contaminant should parallel the behavior and performance of the asphalt itself. Detailed chemical tests have been performed on asphalt liners for disposal sites for uranium mill tailings and for land disposal of radioactive waste. These studies can be used to make a preliminary evaluation of the use of affected soil for CMA incorporation.

The resistance of asphalt to many reagents at atmospheric temperatures is well documented. Prolonged contact with dilute acidic solutions can result in hardening of the asphalt due to formation of asphaltenes. Nitric acid is very reactive with asphalt, even in dilute solutions, whereas hydrochloric acid does not affect asphalt. Asphalt reaction with sulfuric acid is intermediate. Asphalts are generally more resistant to alkaline solutions than to acidic solutions, a favorable characteristic for asphalt liners. However, alkaline solutions can react to form salts such as sodium naphthenates that form excellent emulsifying agents. Theoretically, this could be a problem for affected soil if contaminants were mobilized in the emulsified solution. However, the emulsification depends on the degree of alkalinity and the diffusion and hydraulic resistances of the asphalt, which are generally extremely low, less than 10^{-12} cm^2/s and 10^{-9} cm/s, respectively. Without further experimental verification, emulsification of an asphalt liner is not expected to be important, nor is it leachable to the extent of releasing hydrocarbon constituents in excess of regulatory limits. The resistance of asphalt to selected chemicals under a variety of conditions is listed by MRM Partnership (1988).

7.7 DURABILITY

Durability is primarily dependent upon the aggregate, upon soil resistance to crushing and abrasion, and upon the asphalt's resistance to weathering and aging. The best indication that asphalt liners and structures have long-term durability and performance is the existence of surviving asphaltic structures from antiquity. Asphalt was in general use in western civilizations from about 2000 B.C. to the first century A.D., when its use was superseded by more economic methods of working wood, tar, and pitch, and asphalt deposits were no longer available to existing mining technologies. Surprisingly, the ancient mixes are not that dissimilar to modern ones. Besides the obvious uses as mortars, pavements, revetments, and foundations, at many Mesopotamian sites asphalt liners from 0.1 mm to several centimeters thick were commonly used in drainage, water tanks, and plumbing fixtures. In many of these cases, the asphaltic structures that have escaped intentional or accidental destruction are still adequately performing their primary function. Samples taken from these structures often show favorable properties, including low permeabilities, but it is uncertain how long exposure to UV light and surface conditions has affected the asphalt relative to the general aging processes expected in the subsurface environment.

Durability of asphalt pavement may be enhanced by high asphalt content, compacted impervious mixtures, or application of a protective coating (i.e., chip seal or seal coat). Durability may, however, be adversely impacted by certain chemical reactions. Lighter oil fractions can be removed from the asphalt by oxidation, volatilization, and separation by selective absorption to aggregate particles. Syneresis or staining, the tendency of some asphalts to exude the lighter oil fractions at the surface, may also adversely affect the overall durability of the final product. All these processes have a tendency to alter the composition of the asphalt, causing the asphalt to become hard and brittle, and exposing deeper portions of the asphalt to the elements.

7.8 AGING

Of most importance to an asphalt liner are the effects of aging. The tendency for asphalt to age under atmospheric influences has been known for some time, but not extensively studied since most asphalt structures have had long service despite severe conditions undergone by such structures. Although not documented, hardening and other aging affects might increase mobilization of the petroleum contaminants from the U.S. EPA by supplying pathways out of the asphalt and by causing separation of petroleum constituents from the asphalt phases during aging. However, these effects would have to be excessive and affect a large proportion of the asphalt to mobilize the small amount of contaminants in the 10% fines of the aggregate.

Physical hardening due to peptization, paraffin crystallization, and volatilization occurs to different degrees in all types of asphalt and is unaffected by the presence of petroleum-contaminated soil as a small part of the aggregate. Chemical hardening, however, may be important. The rate of reaction of asphalt with oxygen is very temperature dependent and varies with asphalt type at high temperatures, but below 50°C the reactions are independent of temperature and asphalt type, are restricted to the asphalt surface, and should not be affected by affected soil. The hardening rate is higher in the presence of light, but because of the dark conditions of the subsurface environment, only the aging reactions that occur in the dark are of importance to an asphalt liner. In the dark, oxygen is bound into SO groups after short aging times and into CO groups after long aging times. Petroleum contaminants are not present in great enough quantities to affect the rate or degree of these reactions. Experiments show that the maximum depth of oxygen penetration is in the range of 2.5 to 5 mm, but the rate of hardening reduces considerably with time.

7.9 BIOLOGICAL RESISTANCE

Microorganisms with the ability to attack hydrocarbons including asphalt are very widely distributed in nature. At least one group, Pseudomonadae contain members that can attack asphalt and are capable of growth under either aerobic or anaerobic conditions. Despite their wide distribution, no documented cases of microbiological degradation causing failure of any structure constructed of asphalt have ever surfaced. Although some minor contribution under optimal conditions may, however, have occurred, such cases have not been of significant concern. The absence of significant asphalt susceptibility to degradation by microorganism is reflected in the rate of degradation. Such activity is so slow that for all intents and purposes, asphaltic products are considered immune to bacteria and fungi. In fact, the recovery of artifacts in good condition from pre-Biblical times is one of the reasons why bitumen has been thought to be immune to microbial attack.

Microorganisms can degrade certain asphaltic components under ideal conditions. Research into microbial degradation of asphalt can be summarized as follows.

- There is no single microorganism that will oxidize all asphaltic components.
- Microbial degradation occurs only at the outermost surfaces.
- The higher the molecular weight of the asphalt component, the more resistant it is to microbial degradation.
- Most soil asphalt-oxidizing microbes grow best at pH 6–8.
- Even under ideal conditions, microbial degradation rates do not exceed 10^{-6} cm/d (0.7 mm penetration per 100 years) and are usually an order of magnitude less.
- Anaerobic degradation is much slower than aerobic degradation.

- Microbial attack is fastest for stream-refined bitumens, followed by air-blown and finally coal tar pitches.
- Microbial inhibitors are ineffective over long time periods.
- Environmental factors (e.g., temperature, pH, state of hydrocarbons, nutrient and oxygen concentrations) have to be perfect for a very long time to result in any noticeable asphalt degradation.

Overall, microbial degradation will be unimportant for all practical purposes.

7.10 PERMEABILITY

Permeability tests have been performed on a variety of liners that were first subjected to aging tests. The permeability obtained for each liner is presented in Chapter 11, Table 11-2. Permeability results generated as part of this testing on five samples of cold-mix asphalt samples incorporating affected soil are presented in Table 7-3. Accelerated aging tests of an asphalt liner at 20°C under oxygen partial pressures of 0.21, 1, and 1.7 atm, with continuous exposure to an acidic leachate at 20°C under varying oxygen partial pressures, have been performed. Solution pH values of 2.5, 2.0, and 1.5 were designated as normal, intermediate, and highly accelerated conditions, respectively. Acidity levels were shown to have an unmeasurable effect on asphalt aging. Permeability was used as a means to measure the immediate effectiveness of the asphalt liner. The permeability appears to be relatively unaffected under these exposure conditions as shown in Figure 7-7.

7.11 LEACHABILITY

A normal asphaltic concrete or CMA paving material is somewhat acceptable as an environmentally safe product, even though there may be volatile organic compound (VOC) emissions during manufacture and placement, notably in regard to the hot-mix process. The question of contamination is not frequently associated with asphaltic concrete even though certain halogenated

Table 7-3 Results of Permeability Testing for Cold-Mix Environmentally Processed Asphalt

Asphalt sample no.	Sample length (cm)	Sample diameter (cm)	Bulk volume (cm^3)	Effective permeability (mD)	Hydraulic conductivity (cm/s)
A-B-1	6.02	5.05	120.58	0.135	1.42×10^{-7}
A-B-2	6.81	5.06	136.94	0.013	1.37×10^{-8}
A-B-5	4.87	4.98	94.86	0.034	3.58×10^{-8}
A-B-7	6.42	5.06	129.10	0.096	1.01×10^{-7}
A-B-10	7.52	5.05	150.62	0.142	1.50×10^{-7}

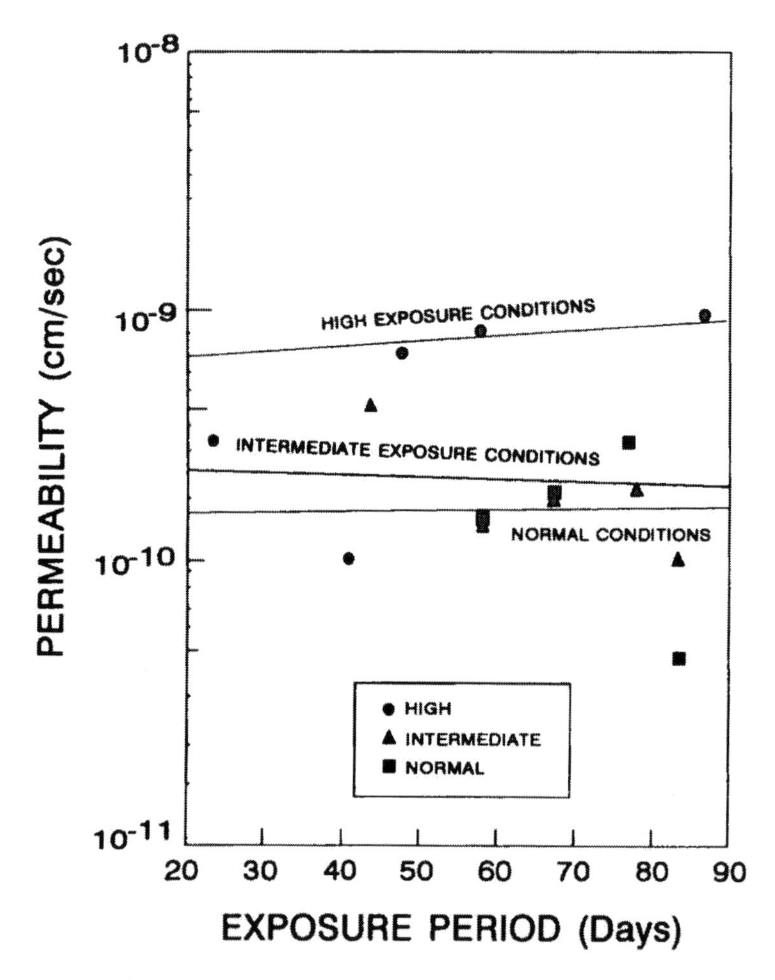

Figure 7-7. Graph showing permeability effects with exposure from asphalt liner.

volatile organics may be present in association with hot-mix bituminous products, notably those containing cut-back asphalt using kerosene, diesel, or other solvents in their dilution process.

Asphalt concrete or CMA may be accepted by Class III landfill and is commonly crushed and used as a component for road-base material. The fact that there are few questions about the safety of asphaltic concrete is justified by the existence of thousands of miles of paved roadways constructed every year. In addition, the performance of leachate tests on asphalt shows a definite lack of contaminants that might leach out under extreme conditions. Leachate analytical results for hazardous waste parameters and certain petroleum constituents in both HMA and CMA incorporating affected soil are presented in Table 7-4.

Table 7-4 Leachate Analysis of Typical Asphaltic Concrete Samples

Parameter	Analytical method	Concentration[a]	Unit	Detection limit
		Cold-mix asphalt		
Cyanide, total	SM412 E/D	ND	mg/kg	1
Flash point	EPA 1010	>220	°F	5
pH	EPA 9045	7.62	—	.01
Sulfide	EPA 376.1	ND	mg/kg	1
Gasoline	EPA 8015	ND	mg/l	.5
Diesel	EPA 418.1	ND	mg/l	.5
		Hot-mix asphalt		
Cyanide, total	SM412 E/D	ND	mg/kg	1
Flash point	EPA 1020	>220	°F	5
pH	EPA 9085	8.43	—	0.01
Sulfide	EPA 376.1	ND	mg/kg	1
Gasoline	EPA 8015	ND	mg/l	.5
Diesel	EPA 418.1	ND	mg/l	.5

[a] ND = not detected above the respective detection limit.

7.12 DISCUSSION

The incorporation of affected soil as an ingredient to produce a commercially viable asphaltic product is a matter of stabilization, solidification, and encapsulation. Asphalt studies of road pavement properties, leaching behavior, sensitivities to moisture-damage, and functional group analysis have provided information that can be used to evaluate the stability of various hydrocarbon compounds- and metals-affected soils that have been utilized as an ingredient to produce an asphaltic end product. These studies indicate that asphalted contaminated soil will be highly stable and perform adequately as viable commercial asphaltic end products. Asphalt is structurally and chemically complex and poorly understood, and is dependent upon the crude petroleum source and any chemical treatment and/or chemical modifiers added during processing. Asphalt is composed of a minimum of ten important functional groups, whose behavior at the asphalt-aggregate interface determine the overall chemical stability of the asphalt. Thermodynamics of the functional groups for asphalt are basically unknown. In addition, no data exist that correlate leachability rates to the presence or absence of specific functional groups, which suggests that the very low permeabilities and diffusivities are the greatest factor in the retention of contaminants in asphalt cements.

For the best chemical performance, asphalt should have high contents of pyridinic, phenolic, and ketone groups, which can be achieved by selectively choosing the source material. If the situation requires special stability or redundancy, small amounts of shale oil and lime can be used as additives.

Situations and conditions that favor the presence of inorganic sulfur, monovalent salts, and high "ionic strength solutions" in the asphalt should be avoided because these conditions potentially decrease the chemical stability of the asphalt cement by disrupting of the functional group-aggregate bonds and increasing the overall permeability. However, these conditions are not expected in the anticipated uses of asphalt cement to fixate or stabilize, solidify, and encapsulate a variety of hydrocarbon- and metals-affected soil using asphalt incorporation technologies.

BIBLIOGRAPHY

Amarantos, S. G. and Petropoulos, J. H., (1981), Certain Aspects of Leaching Kinetics of Solidified Radio Wastes — Laboratory Studies, DEMO 8-1/2, Greek Atomic Energy Commission, Athens, Greece.

Asphalt Institute, 1991, Asphalt Use in Water Environments, Information Series No. 186 (IS-186), 4 pp.

Atlas, R. M., 1981, Microbial Degradation of Petroleum Hydrocarbons, an Environmental Perspective, *Microbiol. Rev.*, pp. 180–209.

Benedetto, A. T., Lottman, L. P., Cratin, P. D., and Ensley, E. K., 1970, Asphalt Adhesion and Interfacial Phenomena, Highway Research Record No. 340, National Research Council Highway Research Board.

Brule, B., Ramond, G., and Such, C., 1986, Relationships Between Composition, Structure and Properties of Road Asphalts, State of Research at the French Public Works Central Laboratory, Transportation Research Record 1096, TRB, National Research Council, pp. 22–24.

Buelt, L. T., 1983, Liner Evaluation for Uranium Mill Tailings, Pacific Northwest Laboratory, Final Report, PHL — 4842.

Conca, J. L. and Testa, S. M., 1992, Chemical aspects of environmentally processed asphalt, in *Proceedings of the International Symposium on Asphaltene Particles in Fossil Fuel Exploration, Recovery, Refining and Production Processes*, in press.

Conca, J. L. and Wright, J. V., 1991, Aqueous diffusion coefficients in unsaturated materials, scientific basis for nuclear waste management XIV, in Proceedings on Materials Research Society Symposium Proceedings, Vol. 212, pp. 879–884.

Daiev, Ch. T. and Vassilev, G. P., 1985, On the diffusion of ^{90}Sr from radioactive waste bituminized by the mould method, *J. Nucl. Mat.*, Vol. 127, pp. 132–136.

Daugherty, D. R., Pietrzak, R. F., Fuhrmann, M., and Columbo, P., 1988, An Experimental Survey of the Factors that Affect Leaching from Low-Level Radioactive Waste Forms, BNL-52125, Brookhaven National Laboratory, Upton, New York.

Eschrich, H., 1980, Properties and Long-Term Behavior of Bitumen and Radioactive Waste-Bitumen Mixtures, Swedish Nuclear Fuel and Waste Management Company, Stockholm, Sweden, SKBF KBS Technical Report 80-14, 178 pp.

Fish, W. and Elovitz, M. S., 1990, Cr(VI) Reduction by phenols in immiscible two-phase systems, implications for subsurface chromate transport, *Trans. Am. Geophys. Union*, Vol. 71, p. 1719.

Fuhrmann, M., Pietrazak, R. F., Franz, E. M., Heiser, J. H., III, and Columbo, P., 1989, Optimization of the Factors that Accelerate Leaching, BNL-52204, Brookhaven National Laboratory, Upton, New York.

Harris, J. O., 1958, Preliminary studies on the effects of micro-organisms on the physical properties of asphalt, *Trans. Kan. Acad. Sci.*, Vol. 61, pp. 110–113.

Haxo, J. E., Jr., 1976, Assessing Synthetic and Admixed Materials for Liner Landfills, Gas and Leachate from Landfills, Formation, Collection and Treatment (Edited by E. J. Genetelli and J. Circello), Report No. EPA 600/9-76-004, U.S. EPA, Cincinnati, Ohio, NTIS Report PB 251161, pp. 130–158.

Hickle, R. D., 1976, Impermeable asphalt concrete pond liner, *Civil Eng.*, pp. 56–59.

Jones, T. K., 1965, Effects of bacteria and fungi on asphalt, *Mat. Prot.*, Vol. 4, 39 pp.

Kriech, A. J., 1990, Evaluation of Hot Mix Asphalt for Leachability, Heritage Research Group HRG# 3959A0M3, Indianapolis, IN.

Meegoda, N. J., 1994, Contaminated soils in highway construction, in *Process Engineering for Pollution Control and Waste Minimization* (Edited by D. L. Wise and B. J. Trantolo), Marcel Dekker, New York, pp. 663–684.

Morrison, R. T. and Boyd, R., 1974, *Organic Chemistry*, Allyn and Bacon, Boston.

MRM Partnership, 1988, Bituminous and Asphaltic Membranes for Radioactive Waste Repositories on Land, Report to Department of the Environment, DOE/RW/87.009, Bristol, England.

Nikiforov, A. S., Zakharova, K. P., and Polyakov, A. S., 1987, Physiochemical foundations of bituminization of liquid radioactive wastes from a nuclear power plant with RBMK reactor and the properties of the compounds formed, *Soviet At. Energy*, Vol. 61, pp. 664–668.

Petersen, J. C., 1986, Quantitative Functional Group Analysis of Asphalts Using Differential Infrared Spectrometry and Selective Chemical Reactions — Theory and Application, Transportation Research Record 1096, TRB, National Research Council, pp. 1–11.

Petersen, J. C., Plancher, H., Ensley, E. K., Venable, R. L., and Miyake, G., 1982, Chemistry of Asphalt-Aggregate Interactions, Relationship with Pavement Moisture Damage Prediction Test, Transportation Research Record 843, TRB, National Research Council, pp. 95–104.

Preston, R. L. and Testa, S. M., 1991, Permanent fixation of petroleum-contaminated soils, in Proceedings of the National Research and Development Conference on the Control of Hazardous Materials, Hazardous Materials Control Research Institute, pp. 4–10.

Sposito, G., 1984, *The Surface Chemistry of Soils*, Oxford University Press, New York.

Testa, S. M., 1995, Chemical aspects of cold-mix asphalt incorporating contaminated soil, *J. Soil Contam.*, Vol. 4, No. 2, pp. 191–207.

Testa, S. M. and Conca, J. L., 1993, When contaminated soil meets the road, *Soils-Anal. Monitoring Remediation*, pp. 32–38.

Testa, S. M., Patton, D. L., and Conca, J. L, 1992a, Petroleum-Contaminated Soils via Cold-Mix Asphalt for Use as a Liner, in Proceedings of the National Research and Development Conference on the Control of Hazardous Materials, Hazardous Materials Control Research Institute, Materials Park, OH, pp. 30–33.

Testa, S. M., Patton, D. L., and Conca, J. L., 1992b, The use of environmentally processed asphalt as a contaminated soil remediation method, in *Hydrocarbon Contaminated Soils and Groundwater* (Edited by Kostecki and Calabrese), CRC/Lewis, Boca Raton, FL, pp. 413–426.

Testa, S. M. and Patton, D. L., 1992, Add zinc and lead to pavement recipe, *Soils-Anal. Monitoring Remediation*, pp. 22–35.

Testa, S. M. and Patton, D. L.,1992, Remediation of metal-affected soil via asphaltic metals incorporation, in Proceedings of Superfund '92, Hazardous Material Control Research Institute.

Testa, S. M. and Patton, D. L., 1994a, Soil Remediation via Environmentally Processed Asphalt™ (EPA™), in Proceedings of the TMS Conference on Extraction and Processing for the Treatment and Minimization of Waste, pp. 461–485.

Testa, S. M. and Patton, D. L., 1994b, Soil Remediation with Environmentally Processed Asphalt™ (EPA)™, in *Process Engineering for Pollution Control and Waste Minimization* (Edited by D. L. Wise and B. J. Trantolo), Marcel Dekker, New York, pp. 297–309.

Wolfe, D. L., Armentrout, D., Arends, C. B., Baker, H. M., and Petersen, J. C., 1986, Crude Source Effects on the Chemical, Morphological, and Viscoelastic Properties of Styrene/Butadiene Latex Modified Asphalt Cements, Transportation Research Record 1096, TRB, National Research Council, pp. 12–21.

Yen, T. F., 1990, Asphaltic materials, in *Encyclopedia of Polymer Science and Engineering*, John Wiley and Sons, New York, pp. 1–10.

8 ASPHALTIC EMULSIONS

8.1 INTRODUCTION

Asphalt, the product of the distillation of crude petroleum, can range from hard to brittle to a water-thin liquid. Asphalt cement is made fluid by several means, including heating, addition of a solvent, or emulsification. Emulsions were first developed in the early 1900s although emulsions we are familiar with today were not developed until the 1920s. Emulsions were initially confined largely to spray applications and used as a dust palliative. The overall volume of asphalt cement used has rapidly increased since 1953, whereas the volume of other asphalt products used has remained essentially the same, reflecting the need for high-type HMA mixes.

Emulsions contain two or more liquids which may be virtually insoluble in each other but may be formed into a stable mixture by proper dispersion. Liquid-in-liquid dispersion systems, such as water dispersion in oil or solvents, are known as emulsions. Emulsions can be solvent-based or water-based, the latter being more attractive to reuse and recycling technologies. With solvent-based emulsions, a solvent such as naphtha or kerosene is added to asphalt to enhance fluidity. The end product is referred to as a cutback. In cutbacks, the asphalt cement is essentially in solution. With water-based emulsions, asphalt emulsion is produced when minute particles of broken-up asphalt are dispersed in water with an emulsifier. The tiny droplets of asphalt remain in suspension until the emulsion is used for its intended purpose. In emulsions, the emulsifier is oriented in and around droplets of asphalt cement, which influences dispersion and stable suspension of the asphalt cement in water. The cutback or emulsion reverts to asphalt cement via evaporation (and compaction, which drives the water out and enhances evaporation) of the asphalt carrier, in this case, either the hydrocarbon solvent or the emulsion water.

Asphalt can be softened by three means, heat, use of a cutback, or use of a surfactant. Heat is added to soften the asphalt material to improve the coating ability and produce bituminous concrete. Cutbacks involve the mixing of the asphalt with a liquid solvent. Solvents used include gasoline, kerosene, and diesel fuel to produce a rapid, medium, or slow cutback, respectively. These

Semi-Solid	Softening Agent	Result	Aggregate	Product	Environmental Effect
Asphalt Cement	Heat	Melted Asphalt	Contaminated Soil	Hottop	Release Volatiles
	Gasoline	Rapid Cutback		Rapid Cold Patch	Leach Gasoline
	Kerosene	Medium Cutback		Medium Cold Patch	Leach Kerosene
	Diesel	Slow Cutback		Slow Cold Patch	Leach Diesel
	Surfactant	Emulsified Asphalt		Cold Mix	Leach Water

Figure 8-1. Matrix showing processes to produce various asphaltic products from asphalt cement.

cutbacks, when used for paving products, become what is referred to as rapid, medium, or slow cold patch. When asphalt is emulsified via use of a surfactant as a softening agent, the result is an asphalt emulsion which produces a cold-mix product. Water-based cold-mix asphalt emulsions are by far the most desirable product to be utilized as a binding agent for environmental purposes. A matrix illustrating these various asphalt processes is presented in Figure 8-1.

The primary uses of emulsions in the United States are for stabilization, surface treatments, patching and overlays, slurry sealing, base and surface coarse mixes, and recycling. What makes a certain emulsion appropriate for a particular project is determined in part on its ability to adequately coat the aggregate being used. Factors of importance in this consideration include aggregate type, gradation and characteristics of the fines, anticipated water content of the aggregate, and availability of water at the project site.

Presented in this chapter is a discussion of emulsion chemistry, emulsion production, specifications, and breaking and curing. Also presented is a discussion of the various emulsion tests and considerations for emulsion selection.

8.2 EMULSION CHEMISTRY

Emulsions are made up of asphalt, water, and an emulsifying agent, which may contain a stabilizer. Asphalt cement makes up about 55 to 70% of the emulsion. Most emulsions are produced with hardness values of the base asphalt cement in the range of 100 to 250 penetration. Climatic conditions may influence whether a harder or softer base asphalt is used. Asphalt itself is a colloid composed primarily of asphaltenes and maltenes. The asphaltenes and maltenes are the dispersed and continuous phases, respectively. The asphaltenes are believed to furnish hardness, whereas the maltenes provide adhesive and ductile properties to the asphalt and have an influence on viscosity (flow properties). The actual colloidal makeup is a function of the chemical nature of these fractions and their relationship to each other. In addition, the complex interactions of the different fractions makes it difficult to predict the overall behavior of an asphalt to be emulsified; thus, high quality control requirements are necessary during production.

Water contributes significantly to the desired properties of the finished product by wetting, dissolving certain constituents, adhering to other substances, and moderating chemical reactions. Water containing certain impurities, notably positive or negative ions either in solution or as colloidal suspension, may adversely affect the quality of the emulsion, causing an imbalance in the emulsion components affecting overall performance or premature breaking.

The ultimate properties of the asphalt emulsion depend greatly on the chemical used as the emulsifier. The chemical used, referred to as a surface-active agent or surfactant, determines whether the emulsion is to be classified as anionic, cationic, or nonionic. The purpose of the emulsifier is to keep the asphalt droplets in stable suspension and control the breaking time by changing the surface tension at the interface (i.e., the area of contact between the asphalt droplets and the water). There are numerous chemical emulsifiers available, and care must be taken in choosing the appropriate emulsifier compatible with the asphalt cement being used. For anionic emulsifiers, fatty acids, which are wood-product derivatives such as tall oils, resins, and lignins, are used. Anionic emulsifiers are commonly turned into soap or saponified by reacting with sodium hydroxide or potassium hydroxide. Cationic emulsifiers use fatty amines such as diamines, imidazolines, amidoamines, among others. These amines are converted into soap by reacting with an acid, usually hydrochloric acid. Fatty quaternary ammonium salts are also used to produce stable cationic emulsions. These salts are water soluble and thus do not require the addition of an acid to make them so.

8.3 EMULSION PRODUCTION

Although water and oil (or asphalt) are immiscible, the blending of water and asphalt can be accomplished under carefully controlled conditions. The objective in producing an emulsion is to create a dispersion stable enough for pumping, prolonged storage, and mixing, with the ability to break down quickly after contact with aggregate during mixing or after spraying, and to maintain the durability, adhesion, and water-resistance characteristics of the asphalt cement produced.

A typical emulsion manufacturing plant includes a high-speed, high-shear colloid mill, emulsifier solution tank, heated asphalt tank, pumps, and flow-metering gauges. The colloid mill is used to divide the asphalt into tiny droplets. The size of the droplets varies with the type of equipment used; however, droplet sizes can vary to smaller than the diameter of a human hair (i.e., 0.001 to 0.010 mm or 0.00004 to 0.0004 in.).

Processing begins with concurrent streams of molten asphalt cement and treated water pouring into the intake of the colloid mill via pumps. Subjected to intensive shear stress during this initial phase, the heated asphalt cement, when fed into the colloid mill, is divided into tiny droplets. Asphalt particle sizes produced will vary; the size is vital to the production of a stable emulsion. These droplets become dispersed in water in the presence of a surfactant, which allows the asphalt to remain in a suspended state. Since the particles all have a similar electrical charge, they repel each other, which also aids in their remaining in a suspended state.

The newly formed emulsion is then pumped through a heat exchanger to raise the temperature of the incoming emulsifying water (water containing the emulsifying agent) just prior to reaching the colloid mill, although the temperature remains below the boiling point of water to maintain low viscosity. The emulsion is then pumped into bulk storage tanks with some type of stirring apparatus to keep the product uniformly blended. The end result is an attempt to produce a combination of the best features of each component.

8.4 EMULSION SPECIFICATIONS

Standard specifications have been developed by ASTM and the American Association of State Highway and Transportation Officials (AASHTO), as presented in Table 8-1 and Table 8-2. Although not all of these grades are typically available from a particular producer, in most cases they can be made available. For the most part, no solvents are contained in the emulsions (ASTM D 977 and AASHTO M 140), although some cationic emulsions contain a restricted amount of solvent (i.e., ASTM D 2397 and AASHTO M 208).

Asphalt emulsions are divided into three categories, anionic, cationic, and nonionic, reflecting the electrical charges that surround the asphalt particles. Such categories are based on the premise that like charges repel one another, whereas unlike charges attract one another. Thus, when two poles are immersed in a liquid (i.e., an anode and a cathode) and an electrical current is passed through it, the anode becomes positively charged and the cathode becomes negatively charged. Anionic emulsions are generally alkaline, with a pH greater than 7. In anionic emulsions, negatively charged particles of asphalt will migrate to the anode when an electrical current is passed through the emulsion. Conversely, positively charged asphaltic particles will migrate to the cathode in a cationic emulsion. Cationic emulsions are generally acidic, with a pH less

Table 8-1 General Specifications for Various Grades of Anionic and Nonionic Emulsions

	Medium-setting								Slow-setting			
	HFMS-1		HFMS-2		HFMS-2h		HFMS-2s		SS-1		SS-1h	
	Min	Max	Min	Max	Min	Max	Min	Max	Min	Max	Min	Max
Tests on emulsions												
Viscosity, Saybolt Furol at 77°F (25°C), s	20	100	100	—	100	—	50	—	20	100	20	100
Viscosity, Saybolt Furol at 122°F (50°C), s	—	—	—	—	—	—	—	—	—	—	—	
Storage stability test, 24-h, %	—	1	—	1	—	1	—	1	—	1	—	1
Demulsibility, 35 ml, 0.02 N $CaCl_2$, %	—	—	—	—	—	—	—	—	—	—	—	—
Coating ability and water resistance												
Coating, dry aggregate	good		good		good		good		—		—	
Coating, after spraying	fair		fair		fair		fair		—		—	
Coating, wet aggregate	fair		fair		fair		fair		—		—	
Coating, after spraying	fair		fair		fair		fair		—		—	
Cement mixing test, %	—	—	—	—	—	—	—	—	—	2.0	—	2.0
Sieve test, %	—	0.10	—	0.10	—	0.10	—	0.10	—	0.10	—	0.10
Residue by distillation, %	55	—	65	—	65	—	65	—	57	—	57	—
Oil distillate by volume of emulsion, %	—	—	—	—	—	—	1	7	—	—	—	—
Tests on residue from distillation test												
Penetration, 77°F (25°C), 100 g, 5 s	100	200	100	200	40	90	200	—	100	200	40	90
Ductility, 77°F (25°C), 5 cm/min, cm	40	—	40	—	40	—	40	—	40	—	40	—
Solubility in trichloroethylene, %	97.5	—	97.5	—	97.5	—	97.5	—	97.5	—	97.5	—
Float test, 140°F (60°C), s	1200	—	1200	—	1200	—	1200	—	—	—	—	—

Table 8-1 General Specifications for Various Grades of Anionic and Nonionic Emulsions *(continued)*

	Medium-setting								Slow-setting	
	RS-1		RS-2		MS-1		MS-2		MS-2h	
	Min	Max	Min	Max	Min	Max	Min	Max	Min	Max
Tests on emulsions										
Viscosity, Saybolt Furol at 77°F (25°C), s	20	100	—	—	20	100	100	—	100	—
Viscosity, Saybolt Furol at 122°F (50°C), s	—	—	75	400	—	—	—	—	—	—
Storage stability test, 24-h, %	—	1	—	1	—	1	—	1	—	1
Demulsibility, 35 ml, 0.02 N $CaCl_2$, %	60	—	60	—	—	—	—	—	—	—
Coating ability and water resistance										
Coating, dry aggregate	—		—		good		good		good	
Coating, after spraying	—		—		fair		fair		fair	
Coating, wet aggregate	—		—		fair		fair		fair	
Coating, after spraying	—		—		fair		fair		fair	
Cement mixing test, %	—	—	—	—	—	—	—	—	—	—
Sieve test, %	—	0.10	—	0.10	—	0.10	—	0.10	—	0.10
Residue by distillation, %	55	—	63	—	55	—	65	—	65	—
Oil distillate by volume of emulsion, %	—	—	—	—	—	—	—	—	—	—
Tests on residue from distillation test										
Penetration, 77°F (25°C), 100 g, 5 s	100	200	100	200	100	200	100	200	40	90
Solubility in trichloroethylene, %	97.5	—	97.5	—	97.5	—	97.5	—	97.5	—
Float test, 140°F (60°C), s	—	—	—	—	—	—	—	—	—	—

Note: The American Society for Testing and Materials takes no position respecting the validity of any patent rights asserted in connection with any item mentioned in this standard. Users of this standard are expressly advised that determination of the validity of any such patent rights, and the risk of infringement of such rights, is entirely their own responsibility.

Table 8-2 General Specifications for Various Grades of Cationic Emulsions

	Rapid-setting				Medium-setting				Slow-setting			
	CRS-1		CRS-2		CMS-2		CMS-2h		CSS-1		CSS-1h	
	Min	Max	Min	Max	Min	Max	Min	Max	Min	Max	Min	Max
Tests on emulsions												
Viscosity, Saybolt Furol at 77°F (25°C), s									20	100	20	100
Viscosity, Saybolt Furol at 122°F (50°C), s	20	100	100	400	50	450	50	450				
Storage stability test, 24-h, %		1		1		1		1		1		1
Classification test	passes		passes									
Coating ability and water resistance												
Coating, dry aggregate					good		good					
Coating, after spraying					fair		fair					
Coating, wet aggregate					fair		fair					
Coating, after spraying					fair		fair					
Particle charge test	positive		positive		positive		positive		positive		positive	
Sieve test, %		0.10		0.10		0.10		0.10		0.10		0.10
Cement mixing test, %										2.0		2.0
Distillation												
Oil distillate, by volume of emulsion, %		3		3		12		12				
Residue, %	60		65		65		65		57		57	
Tests on residue from distillation test												
Penetration, 77°F (25°C), 100 g, 5 s	100	250	100	250	100	250	40	90	100	250	40	90
Ductility, 77°F (25°C), 5 cm/min, cm	40		40		40		40		40		40	
Solubility in trichloroethylene, %	97.5		97.5		97.5		97.5		97.5		97.5	

than 7. Nonionic emulsions are neutral; therefore, the asphalt particles do not migrate to either pole.

Anionic and cationic emulsions are commonly used in roadway construction and maintenance. Nonionic emulsions are not commonly used, although their use should increase with advances in emulsion technology.

Emulsions can be further classified based on how quickly the asphalt will set (i.e., coalesce and revert to asphalt cement). Under this characteristic, emulsions are classified as either rapid-setting (RS), medium-setting (MS), or slow-setting (SS). The tendency for an emulsion to coalesce is closely related to its ability to mix. For example, SS emulsions are designed to mix with fine-grained aggregates, whereas MS emulsions are designed to mix with fine-grained aggregate. RS emulsions, on the other hand, have little to no ability to mix with an aggregate.

Emulsions preceded by the letter "C" are cationic. The absence of the letter "C" denotes that the emulsion is either an anionic or nonionic emulsion. High-float medium-setting (HFMS) emulsions are anionic and are used primarily in CMA and HMA plant mixes, coarse-aggregate seal coats, and road mixes. These emulsions have a specific quality allowing for a thicker film coating without significant drainage. Quick-set (QS) emulsions are another grade, which was developed for slurry seals. A seldom used grade is an inverted emulsion, which is developed with the water dispersed in asphalt, usually a cutback.

A grade followed by an "h" refers to the use of a harder-base asphalt. HF preceding some of the MS grades denotes high float as measured by the Float Test (ASTM D 139 or AASHTO T 50). An additional cationic sand-mixing grade, which contains more solvent than other cationic grades, is designated as CMS-2s.

8.5 BREAKING AND CURING

The rate at which asphalt globules separate from the water phase, which is required in order for the emulsion to serve as a cementing and waterproofing agent, is called breaking or setting. Optimum results are achieved by controlling the aggregate sizing or adjusting the emulsion formulation to meet the specific requirements of the aggregate. During production of CMA, breaking does not occur until the loosely consolidated product is laid down and compacted, which drives out the water phase, which ultimately evaporates. For surface treatments and seals, breaking occurs upon contact with a foreign substance such as aggregate or pavement surface. With breaking, the asphalt droplets coalesce and produce a continuous film of asphalt on the aggregate or pavement. The rate of breaking varies with the type and concentration of emulsifying agent used, absorption characteristics of the aggregate, emulsion–aggregate mixtures, gradation and surface area (varying surface areas equate to changes in adsorption) of the aggregate, and atmospheric conditions.

Higher absorption rates will tend to accelerate breaking by removing the emulsifying water more rapidly.

Curing is achieved through evaporation of water. Surface and atmospheric conditions are less critical for cationic emulsions, since these emulsions tend to give up their water more easily and faster. The initial deposition of asphalt on the aggregate is an electrochemical phenomenon. Traditionally, anionic emulsions, with a negative charge on the asphalt droplets, perform best with aggregates having mostly positive charges, such as limestone and dolomite. Cationic emulsions, with a positive charge on the asphalt droplets, perform best with aggregate having mostly negatively charged surfaces, such as granitic or siliceous aggregate.

The main bond between the asphalt film and the aggregate occurs after the loss of the emulsifying water. The water is displaced by several means, including evaporation, pressure (i.e., rolling or compaction during lay-down), or by absorption, or a combination of these factors. With the heavily stabilized MS and SS grades used for pavements, aggregate type is of less importance, although slightly moist aggregate facilitates the mixing and coating process. Strength for SS grades depends mainly on dehydration and adsorption, with water removal via either of these mechanisms. Solvent-free CMS and CSS emulsions require that moisture content of the aggregate be at or near optimum for proper mixing and coating. While solvents aid in the mixing and coating process, provisions must be made for the solvent to evaporate for the mixture to properly cure.

8.6 EMULSIFIED ASPHALT TESTS

Testing of emulsified asphalt is performed to (1) measure certain properties related to handling, storage, and field use, (2) control the quality and uniformity of the product during manufacturing and use, and (3) predict and control field performance. Such tests address composition, consistency, stability, examination of residue, rapid setting classification of cationic emulsions, coating, and weight per gallon. The most commonly used tests as applied to emulsions are summarized in Table 8-3 and further discussed below.

8.6.1 Composition Tests

Composition tests include residue by distillation, oil distillate, residue by evaporation, and particle charge tests. Distillation tests are used to determine the relative proportions of asphalt cement and water comprising the emulsion or oil distillates (in percent by volume of the original emulsion sample) when the emulsified asphalt contains such a distillate. Microdistillation tests or residue-by-evaporation tests can also be performed to measure the percentage of asphalt cement in the emulsion. A residue by evaporation test is used to measure the percentage of asphalt cement in the emulsion by

Table 8-3 Summary of Emulsion Tests

Property	Parameter	ASTM method
Composition	Water content	D 244
	Residual and oil distillate by distillation	D 244
	Oil distillate identification by microdistillation	D 244
	Residue by evaporation	D 244
	Particle charge of emulsified asphalts	D 244
Consistency	Viscosity	D 244
Stability	Demulsibility	D 244
	Settlement	D 244
	Cement mixing	D 244
	Sieve	D 244
	Coating	D 244
	Miscibility with water	D 244
	Freezing	D 244
	Coating ability and water resistance	D 244
	Storage stability	D 244
Examination of residue	Specific gravity	D 70 or D 3289
	Ash content	D 128
	Solubility in trichloroethylene	D 2042
	Penetration	D 5
	Ductility	D 113
	Float test	D 139
Rapid setting cationic emulsion classification	Coating for RS grade	D 244
Fluid coating	Coating and water resistancy	D 244
Emulsified asphalt/job aggregate coating	Mixability and coating for SS grade	D 244
Weight per gallon	Unit weight	D 244

simply evaporating the water. The residue derived yields relatively lower penetration and ductility values in comparison to distillation. The resultant residue can also be used for other tests.

A particle charge test is conducted to identify and differentiate cationic from anionic emulsions. This test involves the deposition of asphalt on an electrode. The benefit of such a test is the prevention of the mixing of grades, which could result in breakdown. Positive (anode) and negative (cathode) electrodes are immersed into a sample of emulsion. A controlled direct current electrical source is connected to the electrodes. After a period of time, an appreciable layer of asphalt is deposited on an electrode, depending upon whether the emulsion is anionic or cationic.

8.6.2 Consistency Tests

Consistency or viscosity tests define a fluid's resistance to flow and are used to determine uniformity and whether the emulsion product has the same handling, mixing, and setting characteristics from shipment to shipment. For emulsified asphalts, the Saybolt Furol viscosity test is used, the results of which are reported in Saybolt Furol seconds. Such tests are routinely performed at two testing temperatures covering the normal working range (25 to 50°C or 77 to 122°F, respectively).

Stability testing is used to demonstrate that an emulsion product is capable of being stored without excessive damage or change. Stability can be assessed by conducting several tests including demulsibility, settlement, cement mixing, sieve, miscibility with water, coating ability, water resistance, storage, examination of residue, RS classification for cationic emulsions, field coating, and mass.

Demulsibility is conducted to assist in selecting the proper emulsion grade by differentiating mixing grade emulsion product from rapid-set types. Demulsibility determines the relative rate colloidal asphalt globules in RS grades will break when spread in thin film on soil or aggregate. During testing, a calcium chloride solution and water are thoroughly mixed with the emulsified asphalt, then poured over a sieve to determine how much the asphalt globules coalesce. For RS emulsions, a very weak solution of calcium chloride and water is used. A high degree of demulsibility is an indication of RS emulsion; thus, the asphalt globules are anticipated to break immediately upon contact with an aggregate.

The cement mixing test evaluates the relative rate at which colloidal asphalt globules in SS grades (vs. use of the demulsibility test for RS grade emulsion) will break when spread in thin films on soil, fine-grained soil, and dusty aggregates. These grades are normally unaffected by calcium chloride solutions as used in the demulsibility test. The anticipated reaction during conduct of a cement mixing test differs for cationic and nonionic emulsions. Cationic emulsions react to Portland cement because of surface area. Nonionic and notably anionic emulsions react with Portland cement, chemically forming a water-soluble salt.

A settlement test is performed to indicate the emulsion's stability in storage, the tendency of asphalt globules to settle during storage, and as an indicator of whether or not the emulsion is to be stored, in terms of quality. Failure indicates that something is out of balance or wrong in the emulsification process. Settlement tests are typically performed for 1 to 5 d.

The sieve test complements the settlement test and is also used to determine the amount of asphalt in the form of rather large globules that may not have been detected during conduct of the settlement test. Such globules could potentially clog the spraying equipment and prevent thin and uniform coatings of asphalt on the aggregate particles. Miscibility with water determines whether MS or SS grades can be mixed with water. Not applicable to RS grades, this test is also used to indicate whether the emulsion is capable of mixing with or being diluted with water.

The coating ability of an emulsion is important in determining whether an emulsion will thoroughly coat the aggregate, withstand mixing action while remaining as an asphalt film on the aggregate, and resist the washing action of water after completion of mixing. Factors that affect the selection of a particular emulsion for effective coating are aggregate type and gradation, characteristics of fines, moisture content, and availability of water during processing. Usually more than one type of emulsion is acceptable for a given aggregate. Selection should be based upon mixture properties as determined by comparative mix designs and a varying water content. Typically used to identify MS grades suitable for mixing with coarse-graded calcareous aggregate, other aggregates may be used if calcium carbonate is omitted throughout the methane. This test is not adaptable to RS or SS grades. A field coating test is performed on-site to determine the emulsion's ability to coat the aggregate and withstand mixing and water resistance of the emulsion-coated aggregate. A specific classification test is used for RS cationic grades to determine these emulsions' ability to coat the aggregate.

The ability of an emulsion to remain at uniform dispersion during storage is determined by performance of a storage stability test. This test measures the permanence of the dispersion as related to time.

8.6.3 Examination of Residue Tests

Examinations of residual-type tests are formulated to assure the same desirable characteristics in the base asphalt cement as in the residual asphalt after emulsification and coalescence. Such tests include specific gravity, penetration, solubility, ductility, and float test. Specific gravity, although not routinely a specification requirement, can be used to make volume corrections at elevated temperatures and when determining quantities needed. Most asphalt cements fall within a range of about 1.0 to 1.05; thus, they weigh 1.0 to 1.05 times as much as a similar volume of water.

The solubility test measures the purity of the asphalt cement and keeps to a minimum the additions, emulsifications, and fillers used to emulsify the asphalt. Portions of the cement that are found soluble in specified solvents, commonly trichloroethylene, represent the active cementing constituents, such as salts, free carbon or nonorganic contaminants such as clay or finely divided mineral matter.

The penetration test is an empirical test of consistency. Thus, this test is used for some viscosity-based asphalt specifications to preclude use of materials of low penetration.

Ductibility tests are performed to assess flexibility of the asphalt. Ductibility, defined as the ability of asphalt cement to be extended or pulled into a narrow thread, is evaluated by performing a ductibility test. The presence or absence of ductibility is usually more important than the degree of ductibility.

A float test measures the consistency of the material being examined. Such a test is performed on the residue from distillation of HFMS emulsions.

The weight per gallon or mass per liter test is performed to determine the weight or mass of asphalt emulsion on gallon or liter, respectively.

8.6.4 Emulsion Tests for Construction Purposes

From a construction point of view, emulsion tests are also performed to evaluate constancy (uniformity and storage stability), classification, construction characteristics, asphalt durability, and asphalt purity. Such previously discussed tests as they pertain to construction purposes are summarized in Table 8-4.

Table 8-4 Summary of Emulsion Tests for Construction Purposes

Property	Test
Constancy	Residue (uniformity)
	Sieve
	Settlement
	Storage stability
	Freeze–thaw
Classification	Demulsibility
	Particle charge
	pH
Construction characteristics	Consistency
	Pumping stability
	Dehydration
	Cement mixing
	Stone coating
	Water resistance
	Miscibility with water
Durability	Penetration
	Float test
	Residue
	Adhesion
	Ductility
Asphalt purity	Solubility

8.7 EMULSION SELECTION

Selection of the appropriate type and grade of emulsion will depend on the type of construction to take place and the nature of the end product. Each grade of asphalt emulsion is designed for a specific use, as summarized in Table 8-5. Construction issues include whether a plant mix will be used or whether the material will be mixed-in-place. End product usage will also need to be evaluated (i.e., dust abatement, light vs. heavy loads, etc.). Other factors include

- Climatic conditions during construction;
- Soil and aggregate type;
- Construction equipment available;
- Hauling distances;
- Water availability; and
- Environmental considerations.

Prior testing on a smaller pilot scale is recommended to assure that the appropriate emulsion type and grade and quantities of soil and aggregate type(s) being used are appropriate and that anticipated end performance criteria are achieved.

How rapidly an emulsion sets has a bearing on the end asphaltic product's intended use. Rapid-setting (RS) emulsions quickly react with the soil and aggregate and revert from an emulsion state to asphalt. RS emulsions are primarily used for spray applications, chip and sand seals, and surface treatments. Medium-setting (MS) emulsions are generally designed for mixing with coarse aggregate and remain workable for a few minutes. MS emulsions are used extensively in mobile plants. As with the RS emulsions, MS emulsions have high viscosities to prevent runoff. High-float anionic type MS emulsions provide better aggregate coating and asphalt retention under extreme temperature conditions. Slow-setting (SS) emulsions provide the maximum mixing stability and are generally used with fine-grained dense soil and aggregate. Their long workability times allow for adequate mixing. When diluted with water, their low viscosities can be reduced and they can be used for tack coats, fog seals, and dust palliatives. Coalescence of asphalt particles depends on the evaporation of the water. Relatively faster rates can, however, be achieved with addition of Portland cement (or crushed concrete debris) or hydrated lime. The SS emulsions are generally used for fine-grained aggregate-emulsion basis, soil-asphalt stabilization, asphalt surface mixes, and slurry seals.

Regardless of the emulsion grade to be used, good adhesion between the emulsion and the aggregate to be used will be dependent on the electrical surface charges of the asphalt droplets and the aggregate. Favorable adhesion is improved if the charges are different. The predominant aggregate surface charge determined via laboratory testing will decide whether an anionic or cationic emulsion will produce the best results.

Table 8-5 Summary of the General Uses of Emulsified Asphalt

Type of construction	ASTM D977 AASHTO M208						ASTM D2397 AASHTO M 140							
	RS-1	RS-2	MS-1 HFMS-1	MS-2 HFMS-2	MS-2h HFMS-2h	HFMS-2s	SS-1	SS-1h	CRS-1	CRS-2	CMS-2	CMS-2h	CSS-1	CSS-1h
Asphalt-aggregate mixtures														
For pavement bases and surfaces														
Plant mix (hot)	—	—	—	—	X[a]	—	—	—	—	—	—	—	—	—
Plant mix (cold)														
Open-graded aggregate	—	—	—	X	X	—	—	—	—	—	X	X	—	—
Dense-graded aggregate	—	—	—	—	—	X	X	X	—	—	—	—	X	X
Sand	—	—	—	—	—	X	X	X	—	—	—	—	X	X
Mixed-in place														
Open-graded aggregate	—	—	—	X	X	—	—	—	—	—	X	X	—	—
Dense-graded aggregate	—	—	—	—	—	X	X	X	—	—	—	—	X	X
Sand	—	—	—	—	—	X	X	X	—	—	—	—	X	X
Sandy soil	—	—	—	—	—	X	X	X	—	—	—	—	X	X
Slurry seal	—	—	—	—	—	X	X	X	—	—	—	—	X	X
Asphalt-aggregate applications														
Treatments and seals														
Single surface treatment (chip seal)	X	X	—	—	—	—	—	—	X	X	—	—	—	—

Table 8-5 Summary of the General Uses of Emulsified Asphalt *(continued)*

	ASTM D977 AASHTO M208						ASTM D2397 AASHTO M 140							
Type of construction	RS-1	RS-2	MS-1 HFMS-1	MS-2 HFMS-2	MS-2h HFMS-2h	HFMS-2s	SS-1	SS-1h	CRS-1	CRS-2	CMS-2	CMS-2h	CSS-1	CSS-1h
Multiple surface treatment	X	X	—	—	—	—	—	—	X	X	—	—	—	—
Sand seal	X	X	X	—	—	—	—	—	X	X	—	—	—	—
Asphalt applications														
Fog seal	—	—	X[b]	—	—	—	X[c]	X(c)	—	—	—	—	X[c]	X[c]
Prime coat-penetrable surface	—	—	—	X[d]	—	—	X[d]	X[d]	—	—	—	—	X[d]	X[d]
Tack coat	—	—	X[b]	—	—	—	X[c]	X[c]	—	—	—	—	X[c]	X[c]
Dust binder	—	—	—	—	—	—	X[c]	X[c]	—	—	—	—	X[c]	X[c]
Mulch treatment	—	—	—	—	—	—	X[c]	X[c]	—	—	—	—	X[c]	X[c]
Crack filler	—	—	—	—	—	—	X	X	—	—	—	—	X	X
Maintenance mix														
Immediate use	—	—	—	—	—	X	X	X	—	—	—	—	X	X

[a] Grades of emulsion other than FHMS-2h may be used where experience has shown that they give satisfactory performance.

[b] Diluted with water by the manufacturer.

[c] Diluted with water.

[d] Mixed-in-place only.

BIBLIOGRAPHY

Asphalt Institute, 1979, A Basic Asphalt Emulsion Manual, Manual Series No. 19, 2nd ed., College Park, MD, 231 pp.

Chemical Publishing Company, 1943, *Emulsion Technology Theoretical and Applied — Including the Symposium on Technical Aspects of Emulsions*, Chemical Publishing Company, Brooklyn, New York, 290 pp.

9 CHEMICAL ASPECTS OF CEMENTITIOUS PRODUCTS

9.1 INTRODUCTION

Stabilization and solidification techniques are used for the incorporation of various wastes into cementitious materials. These wastes include such materials as fly ash, lime, and granulated blast furnace slag, radioactive waste, and, to a lesser degree, organic and inorganic contaminated soil. The mechanisms involved in solidifying waste into cementitious forms and producing a viable end product are not fully known but can be described in terms of mechanical stabilization and immobilization by fixation and isolation. These characteristics are achieved by a well-designed hydraulic cement and/or fly ash stabilization process. Upon hardening, the mass is bonded and strengthened, the contaminants are coated and incorporated in the siliceous material, and connectivity between pores is significantly reduced. As with asphalt, much is known regarding conventional cement technology, although not until recently has there been much attention focused on the incorporation of hazardous, toxic, and radioactive constituents in a cementitious matrix, specifically in regard to the specific mechanisms involved. As with any reuse and recycling approach, however, designing a suitable form for a given waste and predicting anticipated performance over decades or even centuries, as when incorporating low- and intermediate-level radioactive waste, are challenging.

Presented in this chapter is a discussion of types of cements, pertinent physical properties in regard to setting time, comprehensive and flexural strength and durability, and pertinent chemical properties such as general chemistry, pH, internal redox potential, and sorption potential. Leachability and degradation as they relate to reuse and recycling is also discussed.

9.2 TYPES OF CEMENT

The development of cement over the ages has evolved from lime mortar to natural cement to what we commonly refer to today as Portland cement.

There are presently many types of cement products as described by ASTM, the Canadian Standards Association (CSA), and others. Other conventional and common types, and their general usage are summarized in Table 9-1.

Lime has been used as a binder since recorded history. Limestone ($CaCO_3$), when heated to approximately 850 to 900°C, decomposes to CaO or lime, and gaseous carbon dioxide (CO_2). The produced lime is powdery in nature and loosely coherent. When mixed with water, however, a significant amount of heat is liberated, to form what is referred to as slaked lime $(CaOH)_2$. Slaked lime when mixed with sand produces a weak mortar of high open porosity and moderate solubility. With a solubility of 1.1 g l^{-1} at 18°C, a strongly alkaline saturated solution with a pH of 12.4 results. The open porosity and moderate solubility of lime results in a mortar that is not very desirable and obviously subject to leaching.

9.2.1 Natural Cements

Natural cements were widely used in the United States until the early twentieth century, when they were gradually replaced by Portland cement. Thermally activated limestone contains a proportion of clay. When heated, the clay minerals lose water, forming poorly crystallized, fine-grained anhydrous aluminosilicates, which are then mixed with water and allowed to set, forming a silicate gel. This gel is usually designated as C-S-H (i.e., Calcium or Ca-Silica or Si-Water or H_2O). Dehydroxylated kaoline reacts with slaked lime, resulting in an equation that essentially controls the bonding which occurs in Portland and modified Portland cements and their overall durability:

$$Ca\ (OH)2 + Al_2\ Si_2\ O_7 + H_2O \rightarrow C\text{-}S\text{-}H$$

This equation is, however, incompletely balanced with the C-S-H product having a variable Ca:Si ratio, and Al can react with certain constituents to form other phases not represented in the above equation. The C-S-H, however, does tend to fill pores and enhance bonding and, thus, strength. The C-S-H is also more resistant to leaching in comparison to $Ca\ (OH)_2$ and is considered much superior to lime mortars overall.

9.2.2 Portland Cement

Portland cement as a hydrated paste fulfills the role as a binder in concrete. When mixed with sand, gravel, and water, Portland cement is commonly used for construction and engineering applications and for stabilization for solidification/reuse/recycling technologies. The composition of Portland cement evolved from that of natural cements. At higher burning temperatures of approximately 1450°C, all the SiO_2 components, such as quartz and clay, react to form a phase assemblage in equilibrium. This phase assemblage consists

Table 9-1 Summary of Cement Types

Cement type	Description	Usage
ASTM Type I (CSA Normal)	Portland cement	General purpose. Solidification/stabilization. Reuse/recycling.
ASTM Type IA	Same as Type I	Improves resistance to freeze–thaw.
	Contains air-entraining agents	Improves resistance to scaling.
ASTM Type II (CSA Moderate)		Where moderate sulfate attack anticipated. Where moderate heat of hydration required.
ASTM Type IIA	Same as Type II	Improves resistance to freeze-thaw.
	Contains air-entraining agents	Improves resistance to scaling.
ASTM Type III (CSA High Early Strength)		Where high early strength is required. Cold weather.
ASTM Type IIIA	Same as Type III	Improves resistance to freeze–thaw.
	Contains air-entraining agents	Improves resistance to scaling.
ASTM Type IV (CSA Low Heat of Hydration)	Low heat of hydration Strength development occurs more slowly in comparison to Type I	Massive structures. Where temperature rise must be controlled.
ASTM Type V (CSA Sulfate Resisting)	Develops strength slowly	High sulfate content in soil or groundwater.
ASTM Type IS	Portland/blast furnace slag cement	Commonly used in Europe.
	Low early strength Hardens slowly similar to Type I	General purpose.
ASTM Type IS-A	Contains air-entraining agents	Improves resistance to freeze–thaw. Improves resistance to scaling.
ASTM Type IP	Pozzolan-containing cement Similar to Type IS Low early strength Hardens slowly similar to Type I	General purpose.

Table 9-1 Summary of Cement Types *(continued)*

Cement type	Description	Usage
ASTM Type P	Contains air-entraining agents	Improves resistance to freeze-thaw. Improves resistance to scaling.
Expanding cement	Contains slag Contains calcium sulfoaluminate cement[a] Expands slightly upon hydration (formation of ettriugite)	Where expansion required.
High-aluminum cement	Fused limestone and bauxite with small amounts of silicon and titania Sets slowly High strength develops quickly Potential long-term stability Not a Portland cement	Where quick high strength but slow setting required. Resource availability.
Masonry cement	Contains one or more hydrated lime, limestone, chalk, talc, shale, slag, or clay Favorable workability Favorable plasticity Favorable water restriction	Mortar in bonding brick and masonry.
Natural cement	Similar to Portland cement but produced at lower temperatures below sintering point Not restricted to magnesium content	Mortar.
Sonel cement (magnesium oxychloride cement)	Contains MgO added to solution of magnesium oxide Hard Not water resistant Not a Portland cement	Hardness required.
Waterproofed cement	Produced by intergrinding small amounts of calcium, aluminum or other stearates with cement clinker	Waterproofing required.

[a] Slag reacts in presence of $Ca(OH)_2$ and gypsum in the cement parts.

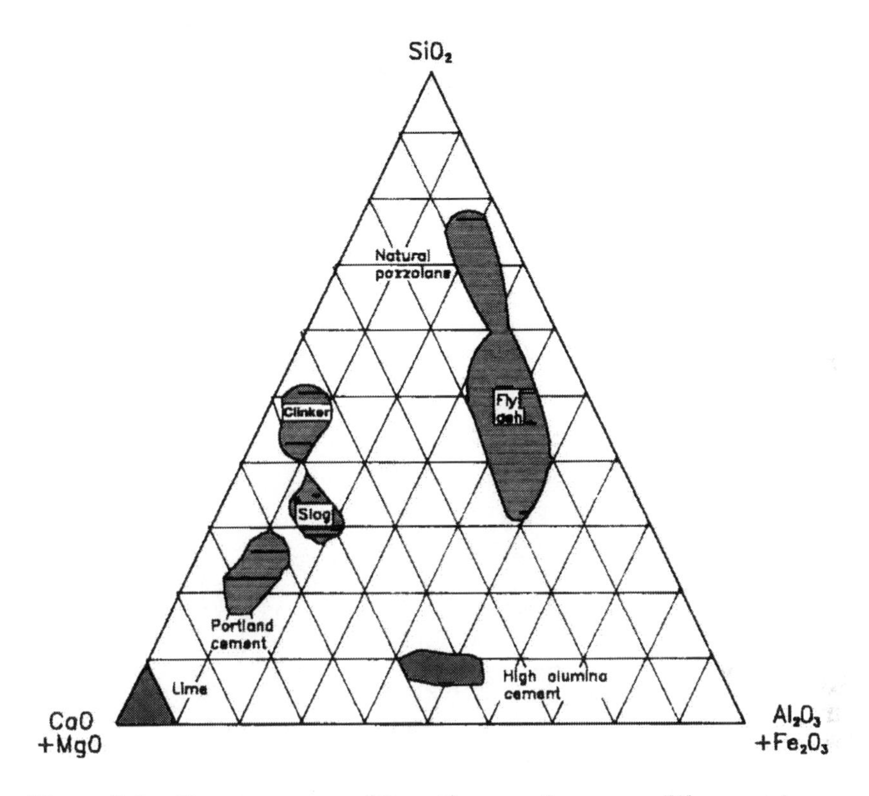

Figure 9-1. Ternary composition diagram for cementitious systems. (After Conner, J. R., 1990.)

of two anhydrous calcium silicates (Ca_3SiO_5 and Ca_2SiO_4), and calcium aluminate and calcium aluminoferrite (Al_2O_3 and Fe_2O_3). A ternary composition diagram for this cementitious system is shown in Figure 9-1.

Batching and calcination ensure that silica is chemically combined as Ca_3SiO_5 and Ca_2SiO_4, with negligible gaseous CaO. During calcination, partial melting and sintering produce a product called clinker. Clinker mixed with 2 to 8% gypsum ($CaSO_4$-$2H_2O$) delays setting of the cement for 1 to 2 h, also allowing for the mixture to attain a period of plasticity while being worked. Small quantities of iron ore or silica may also be added to obtain a desired clinker composition. This mixture, ground to a fine powder with a specific surface of 3000 to 4000 cm^2g^{-1}, is referred to as Portland cement. The principal compounds and characteristics of typical Portland cement are presented in Table 9-2.

The most common silicate used in the reuse and recycling of contaminated soil into a cementitious matrix is Portland cement, Type I, and a 38% solution

Table 9-2 Chemical Composition of Portland Cement (Type I) and Sodium Silicate (Grade N)

Substance	Chemical composition	Amounts (%)
Sodium silicate (Grade N)	SiO_2	28.7
	N_2O	8.90
	Solids	37.6
	Specific gravity	1.39
	Density	11.6 lb/gal
	Weight ratio (SiO_2/Na_2O)	3.22
	pH	11.3
Portland cement (Type I)	$3CaO \cdot SiO_2$	4.5
	$2CaO \cdot SiO_2$	27
	$3CaO \cdot Al_2O_3$	11
	$3CaO \cdot Al_2O_3 \cdot Fe_2O_3$	8
	MgO	2.9
	$CaSO_4$	3.1
	Free CaO	0.5
	Silica-to-sodium oxide ratio	3.22

Modified from Conner, J. R., 1990, *Chemical Fixation and Solidification of Hazardous Waste,* van Nostrand Reinhold, New York, 692 pp.

of sodium silicate. Sodium silicate, commonly known as water gel, is a viscous liquid similar to syrup, with a density and pH of 11.6 lb/gal and 11.3, respectively. Portland cement serves as an agent, whereas the sodium silicate serves as an accelerator or anti-inhibitor for the setting of the cement. The general chemical composition of these two materials of typical Portland cement (Type I) and sodium silicate (Grade N) is presented in Table 9-3.

9.3 PHYSICAL PROPERTIES

Overall performance of a cementitious matrix incorporating contaminated soil is typically judged based on setting time, strength, and durability. These physical properties are described below.

9.3.1 Setting Time

The hardening or setting time of concrete depends upon reactions that take place between the cement and water, as defined by ASTM method C-403. The test consists of measuring the force required to cause a needle to penetrate 1 in. into the mortar. Initial and final setting times are defined as the times penetration resistances reach 500 and 4000 psi, respectively. These values do

Table 9-3 Principal Compounds and Characteristics of Portland Cement

Factor	Principal compounds			
	$CaO \cdot S_1O_2$	$B \cdot CaO \cdot S_1O_2$	$CaO \cdot Al_2O_3$	$CaO \cdot Al_2O_3Fe_2O_3$
Abbreviated formula	C_3S	βC_2S	C_3A	C_4AF
Common name	Alitc	Belite	—	Ferrite phase, FSS
Principal impurities	MgO	MgO	SiO_2	SiO_2
	Al_2O_3	Al_2O_3	MgO	MgO
	Fe_2O_3	Fe_2O_3	Alkalis	
Common crystalline form	Monoclinic	Monoclinic	Cubic orthorhombic	Orthorhombic
Amount present in cement (average %)	50 (35–65)	25 (10–40)	8 (0–15)	8 (5–15)
Reaction rate with water	Medium	Slow	Fast	Medium
Heat of hydration (cal/g)	120 (medium)	60 (low)	320 (high)	100 (medium)
Contribution strength				
Early age	Good	Low	Good	Good
Ultimate	Good	High	Medium	Medium

Modified from Ezelden, A. S. and Korfiatis, G. P., 1994, in *Process Engineering for Pollution Control and Waste Minimization* (Edited by D. L. Wise and D. J. Trantolo), Marcel Dekker, New York, pp. 271–295.

not actually measure strength, but rather the initial value is representative of the time at which fresh concrete can no longer be properly mixed or worked, whereas the final value reflects the time after which strength begins to develop at a significant rate. Set times are determined using a rate of solidification curve, which is illustrated on a graph of penetration resistance in psi vs. time in hours.

Portland cement, as well as other materials such as fly ash, kiln dust, and lime, all undergo hydration reactions as part of the curing process. Setting time can, however, be retarded with the addition of small amounts of organics (e.g., diesel fuel). Initial and final setting times have been reported to be greater with the incorporation of petroleum-contaminated soil of variable grain size, although such increases are not considered significant. The reason for this is inferred to depend on the hydrophobic nature of the organic contaminant that absorbs onto the crystal faces of the pozzolonic particles, thus effectively inhibiting or blocking the infusion of water and subsequent hydration. The overall compatibility of organic and inorganic waste types with Portland cement (Types I, II, and V) and their effects on setting times are summarized in Tables 9-4 and 9-5, respectively.

Additives such as carbon and clays have been used to improve the retention of hydrocarbons within a solidified cementitious matrix. Efforts toward microencapsulation of organics and hydrocarbons have been successful. Microencapsulation is accomplished in two phases. The initial phase involves the emulsification of the hydrocarbon or organic contaminant. The second phase involves the addition of a water-based silicate that is slightly alkaline (pH of 9.5).

9.3.2 Comprehensive and Flexural Strength

Depending on the end use, the unconfined compressive strength of cementitious material will vary. ASTM C-39 is commonly used to evaluate strength. For burial purposes, only 20 psi is usually required, whereas for construction purposes, the required strength may range up to 4000 psi.

The effects of petroleum-contaminated soil used in cement on compressive and flexural strength can be evaluated in terms of the quantity of soil being incorporated, contaminant type and concentration, and soil type. Compressive and flexural strength of concrete generally decreases upon incorporation of relatively higher petroleum-contaminated soil-to-sand replacement ratios, irrespective of soil type. This adverse effect is due in part to increased separation of the cement particles from the water; thus lesser amounts of cement are available to react with water to produce the hardened binder.

Relatively higher petroleum contaminant quantities also result in lower compressive and flexural strength of the hardened cement. This effect is not as apparent with fine-grained soil; this is attributable to physicochemical effects and the repulsive and attractive interactive forces characteristic of

Table 9-4 Compatibility of Portland Cement with Organic Compounds

Chemical group	Portland cement Type I	Portland cement Type II and V
Alcohols and glycols	Durability: decrease (destructive action occurs over a long time period)	Durability: decrease (destructive action occurs over a long time period)
Aldehydes and ketones	NA	NA
Aliphatic and aromatic hydrocarbons	Set time: increase (lengthen or prevent from setting Durability: no significant effect	Set time: increase (lengthen or prevent from setting)
Amides and amines	NA	NA
Chlorinated hydrocarbons	Set time: increase (lengthen or prevent from setting) Durability: decrease (destructive action occurs over a long time period)	Set time: increase (lengthen or prevent from setting) Durability: decrease (destructive action occurs over a long time period)
Ethers and epoxides	NA	NA
Heterocyclics	NA	NA
Nitriles	NA	NA
Organic acids and acid chlroides	Set time: no significant effect Durability: decrease (destructive action occurs over a long time period)	Set time: no significant effect Durability: decrease (destructive action occurs over a long time period)
Organometallics	NA	NA
Phenols	Set time: no significant effect Durability: decrease (destructive action occurs over a long time period)	NA
Organic Esters	NA	NA

Note: NA = not available.

Modified from Spooner, P. A., et al., 1984, Compatibility of Grouts with Hazardous Waste, EPA Report No. EPA-600/2-84-015, Municipal Environmental Research Laboratory, Cincinnati, OH.

Table 9-5 Compatibility of Portland Cement with Inorganic Compounds

Chemical group	Portland cement	
	Type I	Type II and V
Heavy metal salts and complexes	Set time: increase (lengthen or prevent from setting) Durability: decrease (destructive action begins within a short time period)	Set time: increase (lengthen or prevent from setting) Durability: no significant effect
Inorganic acids	Set time: no significant effect Durability: decrease (destructive action occurs over a long time period)	Set time: no significant effect Durability: no significant effect
Inorganic bases	Set time: no significant effect Durability: no significant effect	KOH and HaOH — Set time: no significant effect Durability: decrease (destructive action occurs over a long time period) Others — Set time: no significant effect Durability: no significant effect
Inorganic salts	Set time: increase (lengthen or prevent from setting) Durability: decrease (destructive action begins within a short time period)	Set time: increase (lengthen or prevent from setting) Durability: no significant effect

Modified from Spooner, P. A., et al., 1984, Compatibility of Grouts with Hazardous Waste, EPA Report No. EPA-600-2-84-015, Municipal Environmental Research Laboratory, Cincinnati, OH.

clayey soil. These forces (i.e., Van der Waals and ionic forces) apparently limit the dispersion of soil particles in the matrix.

Soil type can also affect the compressive and flexural strength of concrete. Fine-grained soil utilized to replace coarser-grained soil, such as sand, may result in mixing and soil preparation difficulties and a decrease in strength.

9.3.3 Durability

Durability of cementitious materials depends upon the presence of fluids and their ability to be transported to and from the cementitious matrix. Thus, to enhance durability, such fluids must be prevented or inhibited from entering or leaving a cementitious matrix. This is achieved in part by maintaining low porosity or via sufficient pore size distribution and also by controlling sufficient component size distribution, rate of hydraulic reactivity, and interactions with other components. All these factors assist in maintaining a low permeability and leach resistant matrix.

Durability testing from a conventional approach is performed if the end-product is expected to be exposed to freeze–thaw and wet–dry cycles. For moderate exposure, freeze–thaw and wet–dry durability is evaluated using ASTM D-560 and ASTM D-554, respectively; for harsh exposure, ASTM C-666 is utilized. Preliminary experimental results indicate that the overall integrity of concrete is not significantly affected by the incorporation of moderate amounts of petroleum-contaminated soil. Results of wet–dry, and freeze–thaw tests show minimal weight loss (i.e., less than 2%), with no observed visible cracking or surface deterioration. The compatibility of organic and inorganic waste types with Portland cement (Types I, II, and V) in regard to durability is summarized in Tables 9-4 and 9-5, respectively.

9.4 CHEMICAL PROPERTIES

Portland cement is essentially a calcium silicate mixture containing predominantly tricalcium (C_3S) and dicalcium silicates (C_2S), with smaller amounts of tricalcium aluminate (C_3A) and a calcium aluminaferrite (C_4AF). Typical weight proportions for typical Type I Portland cement are

C_3S	50%
C_2S	25%
C_3A	10%
C_4AF	10%
Other oxides	5%

The five individual oxides represented as three oxide combinations for cement are SiO_2, CaO and MgO, and Al_2O_3 and Fe_2O_3. A ternary phase diagram with the three primary oxide combinations for various reagents used in solidification and stabilization techniques is shown in Figure 9-2. Although the reactions of the various cement and pozzolanic processes differ, all these reagents have generally the same active ingredients in regard to solidification and stabilization processes. All the reagents with the exception of fly ash originate with natural limestone and clays.

The formation or setting of cement as previously discussed occurs through a series of complex hydration reactions. The overall progression of hydration

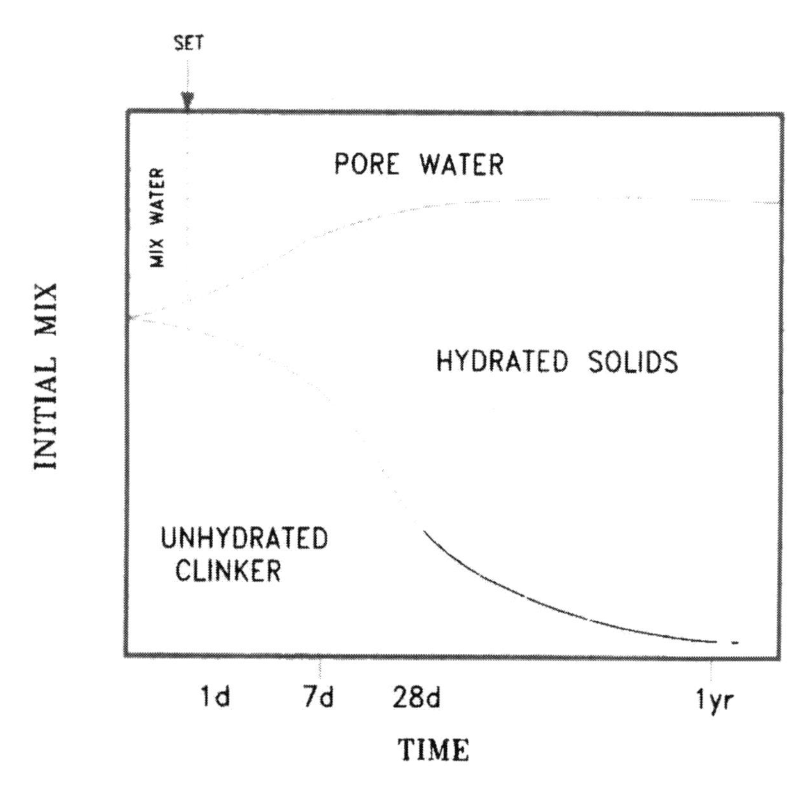

Figure 9-2. Schematic illustrating overall course of cement hydration showing transition from mix water to pore water (dashed line) with time. (After Glasser, F. P., 1993.)

initiates with a burst of heat, during which little hydration occurs, followed by an induction period to retain plasticity (Figure 9-2). Beyond 6 to 24 h, a calcium silicate gel develops. This gel provides strength and good space-filling properties. Between initiation and approximately at 5 to 30°C, hydration reactions accelerate until about two-thirds of the cement have hydrated. Beyond 28 d, hydration continues at a diminishing rate, providing moisture is conserved. By 1 year, 95 to 98% of the cement will have hydrated; however, most cementitious and pozzolanic reactions continue for extremely long periods of time. For all practical purposes, full curing time is assumed to range from 10 to 100 d, commonly averaging 28 d.

Two basic models exist to describe how cement combines with water, the crystalline model and the asmotic or gel model. With both models, the basic reactions occur, but the mechanisms are different. With the crystalline model, the major crystalline compounds hydrate in the presence of water, resulting in different products. Sulfates and C_3A react to form hydrates. If sulfate is abundant, the reaction product is hydrated calcium aluminate sulfate. The calcium aluminate sulfate has the ability to coat the particle surfaces, which prevents rapid hydration. This is exemplified in cements with gypsum. When gypsum is present, setting is retarded; whereas when gypsum is absent, calcium aluminum hydrates form immediately and the system rapidly sets. With the asmotic or gel model, hydration and strength development are due to tricalcium silicate (C_3S) and dicalcium silicate (C_2S), both of which result in the same reaction products, $Ca(OH)_2$ and calcium silicate hydrate gel (C-S-H). With the addition of water for cement, C_3A initially hydrates, resulting in rapid setting and producing a rigid structure. The amount of gypsum controls the setting rate. With the incorporation of contaminated soil, cement systems are tolerant of wet material and restricted by permissible water contents, temperature excursions during solidification, and effects on setting time, strength, and durability.

The end result of hydration is the production of two primary products. These products are $Ca(OH)_2$ and C-S-H at relative volumes of 20 to 25% and 60 to 70%, respectively, with 5 to 15% consisting of other solid phases. The structure of the principal hydration product, C-S-H, is relatively amorphous with high specific surface area (Figure 9-3). Cements do contain some crystalline phases, however, which are summarized in Table 9-6.

9.5 LEACHABILITY

Cementitious materials for the incorporation of contaminated soil display certain chemical features. Those chemical features that favor immobilization of certain chemical constituents as contaminants include

- High surface area of C-S-H, which allows both absorption and adsorption of ions;
- High alkalinity, which allows precipitation of insoluble hydroxides;
- Lattice incorporation into crystalline components of set cements;
- Presence of solubility-limiting phases due to development of hydrous silicates, basic calcium-containing salts, etc.

Factors which may be of importance to the immobilization potential and leachability include internal pH, internal redox potential (E_h), and sorption potential, among others. Cement- and pozzolan-based processes are heavily dependent on pH control for containment of metals. To a lesser degree, some metals may also be bound into the silica matrix under low permeability conditions. High pH is desirable since metal hydroxides have minimum solubility within the range of 7.5 to 11. Since different metals do not behave in a similar manner and vary in regard to minimum solubility, whatever optimum pH is deemed appropriate is a compromise based on the metal contaminants of concern.

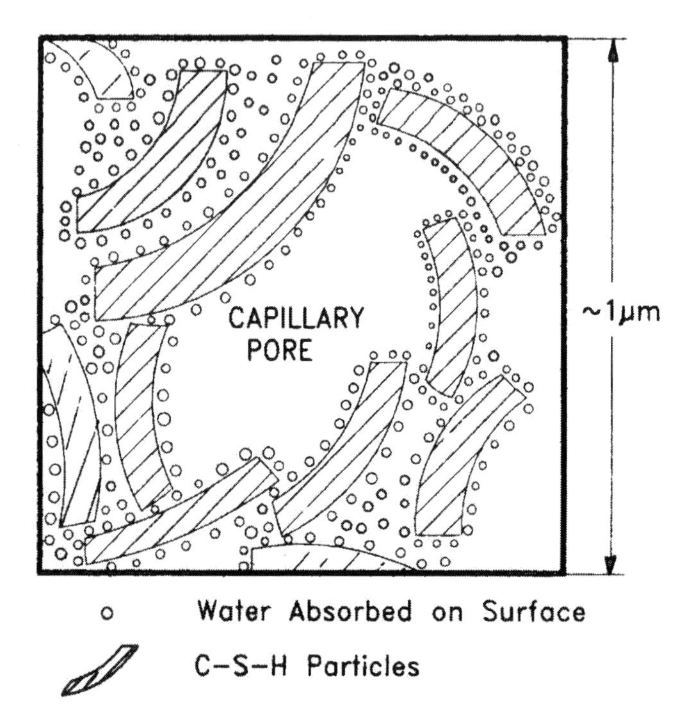

Figure 9-3. Schematic showing the hypothetical reconstruction of the nanostructure of a C-S-H gel. (After Glasser, F. P., 1993.)

Table 9-6 Summary of Crystalline Components of Hydrated Cements

Crystalline form or abbreviation	Chemical formula	Comment
AF_+	$Ca_6Al_2O_6(SO_4)_3 \cdot 32H_2O$	May contain Si substituted for Al, CO^{2-} for SO^{2-}, etc.
Afm	$Ca_4Al_2O_6(SO_4) \cdot 12H_2O$	Forms solid solution
C_4AH_{13}	$Ca_4Al_2O_7 \cdot 13H_2O$	
Hydrogrossulurite or hydrogarnet	$Ca_3Al_{1.2}Fe_{0.8}SiO_{12}H_8a$	Typical composition: Fe not essential
Stratlingite or gehlenite hydrate	$Ca_2Al_2SiO_9 \cdot 8H_2O$	
Portlandite	$Ca(OH)_2$	
Hydrotalcite	$Mg_4Al_2(OH)_{14} \cdot 2H_2O$	Mg/Al ratio variable

Modified from Glasser (1993).

Internal redox potential (E_h) for cement pore fluids is weakly electroactive and typically falls in the range of +100 to +200 mV. The most favorable conditions for cement incorporating wastes are high pH with low E_h. The presence of organics and reactive metals, including steel- or slag-type materials, can be anticipated to lower the redox potential. Materials such as fly ash and slag have different impacts on E_h. Fly ash is composed in part of unburned carbon and chemically reduced iron (i.e., magnetite as Fe_3O_4), which are kinetically too inactive to influence the E_h. Slags are relatively much more electroactive and will tend to decrease the E_h to reflect that of the slag (i.e., in the range of –350 to –450 mV). Adjustments to the E_h can thus be made for controlling speciation and solubility of certain waste constituents.

C-S-H has a high specific surface area with a sorption potential for N_2 on the order of 10 to 50 m^2/g. As schematically illustrated in Figure 9-3, the C-S-H structure is characterized as a layer-structured calcium silicate material with a low state of silicon polymerization. Such structures allow for the availability of strong, unsatisfied surface charges, which cause strong bonding of these layered structures to other C-S-H nanoparticles and water molecules. The stacking of these structures, ranging in size from about 10 to 100 nm, creates large areas of micropores, which can range from a few to a few tens of nanometers in dimension. Overall, the large specific surface area and high density of irregular bonding of water molecules result in a strong sorption potential with variable surface charges ranging from positive to negative values, depending in part on the Ca:Si ratio. For example, during hydration, sodium silicates in conjunction with $Ca(OH_2)$ can tend to result in relatively SO_2-rich C-S-H precipitates, which can have a significant coagulating effect on many metals, taking them out of solution.

Uncertainty remains in regard to the actual location of a hazardous constituent within the cementitious matrix. Several interactions can occur via precipitation, from inclusions, chemisorption, chemical incorporation into the cement structure, formation of a surface compound to any cement component surfaces, or a combination of these situations. These interactions certainly complicate our understanding of the leaching mechanisms involved with various waste types.

Based on leaching experiments using X-ray photoelectron spectroscopy and Fourier transfer infrared and scanning electron microscopy analysis, initial acidic attack is inferred to be on the calcium hydroxide (CH), which occurs as isolated masses and part of tobermorite where the CH is layered between silicate layers. This is followed by acidic attack of the leaching solution on the C-S-H, causing loss of calcium. Silicate polymerization toward silica then occurs, resulting in a solid with less mechanical strength. The resultant decrease in alkalinity also increases solubility of the hazardous action and undermines the overall chemical and physical integrity of the cementitious matrix.

Leachability testing as discussed under Chapter 7 remains the primary tool in determining the maximum contaminant concentration level that can be removed by a liquid from the end product or hardened waste when using a solidification/stabilization technology. TCLP is the most commonly used test. Leachability is dependent on the water-to-cement ratio and the interaction between individual waste components and the cement. Those organic chemical groups that are considered compatible with Portland cement (Types I, II, and V) are presented in Table 9-5.

Some leachability test results have been reported for petroleum-contaminated soil incorporated into cement. Benzene was reported as nondetect after 24 h and 10 d. With metal-contaminated soil, large volume increases ranging from 15% to as much as 150% of the original contaminated soil can occur in order to provide an adequate degree of nonleachability.

9.6 DEGRADATION

Direct attenuation degradation of a cementitious matrix can occur via solid-state diffusion, ingress of aggressive ions, crystallization of phases within the interior of the cementitious matrix, chemical reactions, and leaching. Solid-state diffusion, as with asphalt, is for all practical purposes an extremely slow process. The ingress of ions can occur through the pore structure under the influence of concentration gradients. Crystallization of certain phases within the interior portion of the cementitious matrix can also enhance degradation of the matrix by exerting internal pressure in excess of the matrix's tensile strength. This results in cracking of the cementitious matrix.

Chemical reactions can also occur with the cement parts, C-S-H components, or in other components such as with calcium hydroxide or calcium aluminate hydrates. Sulfate attack can be initiated either from sodium sulfate, calcium sulfate, or magnesium sulfate, the latter being the most aggressive. Degradation as a result of chemical reaction can result in the formation of certain compounds, such as ettringite, brucite, gypsum, and calcite. Leaching of certain ions and species out of the matrix can also result in the overall degradation of the cementitious matrix.

Portlandite ($Ca(OH_2)$) is a primary hydration product of both alite (Ca_3Si0_5) and belite (Ca_2Si0_4). Portlandite has the highest solubility of all the calcium-containing phases in cementitious systems and makes up as much as 25% of the calcium in hydrated Portland cement. Its removal increases connectivity of the cement, allowing accelerated leaching, a decrease in diffusion pathways, and a decrease in strength for each 1% of calcium removal from the matrix. Expansive aggregates and those aggregates well noted for alkali-aggregate reactions should also be avoided.

9.7 BIOLOGICAL RESISTANCE

Most cementitious forms are characterized by high pH and high total alkalinity. Such conditions are not conducive to microbial activity or survival of most microorganisms. This is exemplified by the conventional addition of lime to stabilize sewage sludges. Under oxidizing conditions, biologically enhanced oxidation of insoluble metal sulfides (e.g., pyrite, FeS_2) to soluble sulfates can be of concern. These reactions can result in the release of the metal, which becomes available to generate sulfuric acid, which in turn can attack the cementitious matrix. This can be more prevalent in near-surface environments characterized by low alkalinity.

BIBLIOGRAPHY

Atkins, M. and Glasser, F. P., 1990, Encapsulation of radioiodine in cementitious waste forms, in Scientific Basis for Nuclear Waste Management XIII (Edited by V. N. Oversby and P. W. Brown), Materials Research Society, Pittsburgh, PA, pp. 15–22.

Atkins, M., et al., 1990, The use of silver as a selective precipitant for ^{129}I in radioactive waste management, in *Waste Management* 10, pp. 303–308.

Atkinson, A. and Hearne, J. A., 1984, An Assessment of the Long-Term Durability of Concrete in Radioactive Waste Repositories, U. K. Atomic Energy Authority, Harwell (AERE-R 11465).

Batchelor, B. and Wu, K., 1993, Effects of equilibrium chemistry on leaching of contaminants from stabilized/solidified wastes, in *Chemistry and Microstructure of Solidified Waste Forms*, CRC/Lewis, Boca Raton, FL, pp. 243–259.

Brown, P. W., 1989, Phase equilibrium and cement hydration, in *Materials Science of Concrete* (Edited by J. P. Skalny), American Ceramic Society, Columbus, OH, pp. 73–94.

Butler, L. G., et al., 1993, Microscopic and NMR spectroscopic characterization of cement-solidified hazardous wastes, in *Chemistry and Microstructure of Solidified Waste Forms*, CRC/Lewis, Boca Raton, FL, pp. 151–167.

Cocke, D. L. and Mollah, M. Y. A., 1993, The chemistry and leaching mechanisms of hazardous substances in cementitious solidification/stabilization systems, in *Chemistry and Microstructure of Solidified Waste Forms*, CRC/Lewis, Boca Raton, FL, pp. 187–242.

Conner, J. R., 1990, *Chemical Fixation and Solidification of Hazardous Waste*, Van Nostrand Reinhold, New York, 692 pp.

Conner, J. R., 1993, Chemistry of cementitious solidified/stabilized waste forms, in *Chemistry and Microstructure of Solidified Waste Forms*, CRC/Lewis, Boca Raton, FL, pp. 41–82.

Cote, P., 1986, Contaminant Leaching from Cement-Based Waste Forms under Acidic Conditions, Ph.D. Thesis, McMaster University, Toronto, Ontario, Canada.

Ezeldin, A. S., 1991, Use of coal ash in production of concrete containing contaminated sand, in Proceedings of the American Coal Ash Association 9th International Coal Ash Utilization Symposium, Orlando, FL, pp. 17.1–17.9.

Ezeldin, A. S., Vaccari, D. A., Bradford, L., Dilcer, S., and Farouz, E., 1992, Stabilization and solidification of hydrocarbon-contaminated soils in concrete, *J. Soil Contam.*, Vol. 1, No. 1, pp. 61–79.

Ezeldin, A. S. and Korfiatis, G. P., 1994, Solidification and stabilization techniques for waste control, in *Process Engineering for Pollution Control and Waste Minimization* (Edited by D.L. Wise and D.J. Trantolo), Marcel Dekker, New York, pp. 271–295.

Glasser, F. P., Macphee, D. E., and Lachowski, E. E., 1987, Solubility modelling of cements: implications for radioactive waste immobilization, in *Scientific Basis for Nuclear Waste Management X* (Edited by J. K. Bates and W. B. Seefeld), Materials Research Society, Pittsburgh, PA, pp. 331–341.

Glasser, F. P., 1993, Chemistry of cement-solidified waste forms, in *Chemistry and Microstructure of Solidified Waste Forms*, CRC/Lewis, Boca Raton, FL, pp. 1–39.

Gress, D. L. and El-Korchi, T., 1993, Microstructural characterization of cement-solidified heavy metal wastes, in *Chemistry and Microstructure of Solidified Waste Forms*, Lewis, Boca Raton, FL, pp. 169–185.

Guide to the Disposal of Chemically Stabilized and Solidified Waste, U. S. Environmental Protection Agency, Washington, D. C. (EPA SW-872).

Iler, R. K., 1955, *The Colloid Chemistry of Silica and Silicates*, Cornell University Press, New York, 324 pp.

Inorganic Sulfur Oxidation by Iron Oxidizing Bacteria, 1971, U. S. Government Printing Office, Washington, D. C.

Ivey, D. G., et al., 1993, Electron microscopy characterization techniques for cement solidified/stabilized metal wastes, in *Chemistry and Microstructure of Solidified Waste Forms*, CRC/Lewis, Boca Raton, FL, pp. 123–150.

Macphee, D. E., Atkins, M., and Glasser, F. P., 1989, Phase development and pore fluid chemistry in aging blast furnace slag-portland cement blends, in *Scientific Basis for Nuclear Waste Management XII* (Edited by V. M. Oversby and P. W. Brown), Materials Research Society, Pittsburgh, PA, pp. 117–127.

Martin, J. P., Biehl, F. J., and Robinson, W. T., 1990, Stabilized petroleum waste interaction with silty clay subgrade, in *Petroleum Contaminated Soils* (Edited by E. M. Calabrese and P. T. Kostecki), Vol. 2, CRC/Lewis, Boca Raton, FL, pp. 177–197.

McDowell, T. K., 1992, Remediation of waste motor oil contaminated soil by microencapsulation, in *Hydrocarbon Contaminated Soils*, Vol. II (Edited by P. T. Kostecki, E. J. Calabrese and M. Bonazountas), CRC/Lewis, Boca Raton, FL, pp. 549–558.

McDowell, T. K., 1994, Siallon: the microencapsulation of hydrocarbons within a silica cell, in *Process Engineering for Pollution Control and Waste Minimization* (Edited by D. L. Wise and D. J. Trantolo), Marcel Dekker, New York, pp. 425–439.

Roy, D. M., Grutzeck, M. W., and Mather, K., 1980, PSU/WES Interlaboratory Comparative Methodology Study of an Experimental Cementitious Repository Seal Material, Office of Nuclear Waste Isolation, Battelle Memorial Institute, Columbus, OH (ONWI-198).

Roy, D. M., 1990, Cementitious materials in nuclear waste management, in Cements Research Progress, 1988, American Ceramic Society, Columbus, OH, pp. 262–292.

Roy, D. M. and Scheetz, B. E., 1993, The chemistry of cementitious systems for waste management: the Penn State experience, in *Chemistry and Microstructure of Solidified Waste Forms*, CRC/Lewis, Boca Raton, FL, pp. 83–101.

Salaita, G. N. and Hannak, P. G., 1993, The potential of surface characterization techniques for cementitious waste forms, in *Chemistry and Microstructure of Solidified Waste Forms*, CRC/Lewis, Boca Raton, FL, pp. 103–121.

Scheetz, B. E., et al., 1985, Properties of cement-solidified radioactive waste forms with high levels of loading, *Ceramic Bull.*, Vol. 64, No. 5, pp. 687–690.

Spence, R. D., Ed., 1992, *Chemistry and Microstructure of Solidified Waste Forms*, CRC/Lewis, Boca Raton, FL, 276 pp.

Spooner, P. A., et al., 1984, Compatibility of Grouts with Hazardous Waste, EPA Report No. EPA-600/2-84-015, Municipal Environmental Research Laboratory, Cincinnati, OH.

United States Department of the Interior, 1975, Concrete Manual — A Water Resources Technical Publication, USDI, 8th ed., 627 pp.

Walton, J. C., Plansky, L. E., and Smith, R. W., 1990, Models for Estimation of Service Life of Concrete Barriers in Low-Level Radioactive Waste Disposal, Idaho National Engineering Laboratory, Idaho Falls, ID (NUREG/CR-5542, EGG-2597).

Young, J. F., Berger, R. L., and Lawrence, F. V., 1973, Studies on the hydration of tricalcium silicate pastes. III. Influences of admixtures on hydration and strength development, in *Cement Concrete Res.*, Vol. 3, No. 6.

10 TYPES OF CONTAMINANTS FOR REUSE AND RECYCLING

10.1 INTRODUCTION

The number and volume of materials once considered waste or of little beneficial use that are presently being considered for reuse and recycling have been steadily on the increase. As previously discussed, the majority of these materials have been incorporated into asphaltic pavement and have included rubber, plastic, glass, furnace and steel slag, reclaimed asphalt and concrete, and fly ash. Only during the past few years have materials such as contaminated soil, waste rock, mine tailings, and coal tars also been considered; in many cases these have been successfully utilized for reuse and recycling via incorporation into a variety of asphaltic and cementitious end products. Several of these materials are at the experimental and product development phase or are not developed to the level that large volumes of waste materials can be steadily handled. However, other materials are beginning to be utilized widely as generators understand more fully the cost, effectiveness, aesthetic appeal, and other benefits such reuse and recycling programs offer, including the reduction in the overall liability as a generator to its most negligible form.

Presented in this chapter is discussion of certain contaminants and materials that are either being utilized or being considered for reuse and recycling. Included is a discussion of the suitability of organic and inorganic-contaminated soil, fly ash, coal tar, and low and intermediate radioactive waste.

10.2 ORGANIC-CONTAMINATED SOIL

All hydrocarbon-affected soil, regardless of concentration of the hydrocarbon contaminant of concern, can be considered for reuse and recycling via incorporation into asphaltic and cementitious end products. Petroleum and fuel hydrocarbons, chlorinated hydrocarbons, and monocyclic and polynuclear aromatic hydrocarbons, among others are all suitable candidates for CMA asphalt incorporation. The type, volume, and contaminant concentration of such materials to be reused and recycled, however, is limited with HMA and to other cementitious reuse/recycling technologies.

The basis for the incorporation of hydrocarbon (or organic) affected soil in asphalt depends upon the overmiscibility of various petroleum products. Asphalt is derived from crude oil and is the heavy end of the selective distillation of petroleum, the principal method for separating crude oil into a variety of useful hydrocarbon products. From low to high boiling points, such products range from butanes, gasoline, naphtha, kerosene, light gas-oil, heavy gas-oil, and asphalt and residue, respectively.

Asphalt mixed with other petroleum products is miscible, whereas, when mixed with water, is immiscible. For example, if asphalt is mixed with kerosene, an intermediate oily substance is produced with no development of a separate phase liquid. Thus, any small proportion of petroleum product mixed with asphalt merely produces an asphalt of a slightly different specification or characterization. It is this factor that forms the basis for the incorporation of hydrocarbon-contaminated soil, such as the ubiquitous and commonly encountered fuel hydrocarbons (i.e., gasoline, diesel, etc.) as ingredients in the production of asphaltic products.

This same principle applies to contaminated soil generated at gas manufacturing plants, which are typically affected by MAHs and PAHs. Specific constituents include a minimum of 16 contaminants, such as pyrene, benzo(a) pyrene, fluoranthene, phenanthrene, and others (Table 1-5). Coal tar-contaminated soil has been successfully incorporated into CMA and HMA processes. With CMA, few limitations as to contaminant type and volume exist except that the end product must fulfill its end use. With HMA and coal tar-contaminated soil, some experimental feasibility studies have been performed. Coal tar-contaminated soil has been used to replace reclaimed asphalt pavement during HMA production. These studies have proven successful with coal tar-contaminated soil used in one mix and with contaminated soil stabilized with SS-11 slow-setting asphalt emulsion as an additive in another mix. In these HMA mixes, contaminated soil, aggregate (sand and gravel), and emulsion ranged in volume from 12.4 to 30%, 30 to 43%, and 5.2 to 6.3%, respectively.

10.3 INORGANIC-CONTAMINATED SOIL

Metal-contaminated soil includes soil affected by geothermal mine-scale, foundry residue, steel manufacturing slag, crushed battery casings, and auto wrecking yard residue, all of which have been successfully utilized in asphalt, cement materials, and as an admixture with construction fill. Lead is the predominant contaminant in most of these waste types. Foundry residue differs, however, in that it is composed primarily of silica oxides which may range up to 70 to 95% in weight. The remaining fraction is composed of Fe, Al, Mg, Ca, and their oxides, with typically elevated Mn, Ni, Cu, and Zn. Conventional treatment of metal mine drainages produces a sludge consisting principally of metal hydroxides and oxides, which can comprise up to 33%

of the treated water volume. These wastes are also pH sensitive; exposure to mildly acidic conditions can result in the dissolving of certain metal hydroxides and oxides and mobilization of metals.

Inorganic-contaminated soil essentially focuses on metals. Typical metals of primary environmental concern are listed in Table 5-6. Of the 18 metals noted in Table 5-6, lead is one of the most common contaminants encountered at Superfund sites throughout the United States. Lead is also a common contaminant at industrial sites, where such activities as battery breaking and recycling, auto wrecking, oil refining, paint manufacturing, metal molding and casting, ceramic manufacturing, and primary and secondary smelting have taken place. Lead has thus been the primary focus in assessing toxicity potential to humans and the environment. Over the past 20 years, there has been a shift from industrial exposure to environmental exposure. With the decline in the use of lead in gasoline, both in the United States and in other countries, there has been a diminishing concern with airborne lead but correspondingly higher concern regarding lead accumulation in soil. In addition, lead and other metals can accumulate in soil, industrial manufacturing and mining operations and activities, and as a result of old paint. There still remains today considerable debate regarding a value for lead concentration in soil that will protect the public, especially children.

Several technologies are available for the remediation of lead-affected soil, including extraction (i.e., segregation), stabilization and solidification, vitrification, and electrokinetics. Additional remedial alternatives derived from the mining industry include froth-flotation and heap leach extraction. These technologies become increasingly limited when dealing with fine-grained soils and when lead is combined with other waste material such as petroleum hydrocarbons, waste materials with high carbon content, other metals, etc.

The basis for the reuse and recycling of metal-affected soil does not, in comparison to hydrocarbon-affected soil, depend upon miscibility. Instead, the primary factor that allows metal-contaminated soil to be incorporated into asphaltic products is permeability. Essentially, oil and water are immiscible phases and do not mix. Furthermore, asphalt is characterized by negligible permeability and low diffusion coefficients, as discussed under Chapter 7. Thus, regardless of the physical and chemical nature of the metal, the extremely low permeability of asphalt and extremely low diffusion coefficients allow for incorporation of varying concentrations of a variety of metals, providing the asphaltic product meets the engineering criteria set forth for its intended end use.

10.4 FLY ASH

Power plant ash is the sixth most abundant mineral resource in the United States. The electric utility industry in 1975 burned approximately 410 million tons of coal for the production of electrical energy, producing nearly 63 million tons of ash (41 million tons of fly ash and 22 million tons of dry bottom ash

Table 10-1 Average Chemical Composition of Coal Ash

Constituent	Chemical symbol	Amount (%)
Silicon dioxide	SiO_2	45.7
Aluminum oxide	Al_2O_3	26.0
Ferric oxide	Fe_2O_3	17.1
Calcium oxide	Cal	3.8
Sulfur trioxide	SO_3	2.6
Potassium oxide	K_2O	1.5
Titanium oxide	TiO_2	1.2
Magnesium oxide	MgO	1.2
Sodium oxide	Na_2O	0.6
Phosphorus pentoxide	P_2O_5	0.3

and boiler slag). Since most fly ash ends up at landfills, it is estimated that these quantities may represent as much as 20% of the total material that eventually ends up at landfills.

Fly ash is primarily a by-product of coal burning plants. Fly ash is typically characterized as small, solid spheres with diameters ranging from 0.5 to approximately 200 μm, with solid densities ranging from 1.5 to 1.7 g cm^{-3}. Composition varies greatly depending on the origin of the coal, the degree of pulverization before the coal is burned, and the type of boiler unit utilized during coal burning. Average chemical compound concentrations for coal ash are presented in Table 10-1. Class "C" ash is produced from low rank coals, whereas Class "F" ash is produced from bituminous coal. Class "C" ash is characterized by relatively higher concentrations of CaO, MgO, and SO_3, and low concentrations of S_1O_2 and Al_2O_3, in comparison to Class "F" ash. Due to varying operating conditions of coal burning plants, fly ash can vary in size, shape, and composition. Fly ash is composed of cenospheres ranging up to 20% by volume. Cenospheres are spheres of silicate glass filled with nitrogen and carbon dioxide, ranging up from 20 u to 200 u in diameter.

Fly ash can be incorporated into CMA mixtures quite readily. Its use in HMA processes is not, however, well documented. Class "F" ash is typically not incorporated in concrete due to relatively high carbon content. Class "C" ash, on the other hand, is more favorable for incorporation into concrete due to its relatively low carbon content and high CaO content, providing sulfur content is low. If sulfur is contained in unburned carbon instead of being associated with a Ca phase such as gypsum ($CaS0_4 \cdot 2H_2O$), an acceptable material for incorporation into concrete is produced with the removal of the carbon phase.

In product development, consideration must be given to separation of ash components and physical and chemical characterization of such components. Depending on such characteristics, products produced include utilizing the fly ash as filler in plastics, ceramic products (mullite; $3Al_2O_3$ $25iO_2$), activated carbon from the unburned carbon component, and concrete and brick production. Plastics suitable for use of fly ash in lieu of $CaCO_3$ include ABS, PVC, polyesters, polyethylene, polypropylene, nylon, urethane, rubber, epoxy, and polystyrene. Mullite is a refractory ceramic material used in high temperature applications (e.g., furnace linings). In duplication of mullite (72% Al_2O_3 and 28% S_1O_2), additional Al_2O_3 is added to the fly ash. Conversion of a fly ash and Al_2O_3 mixture into a dense mullite product is accomplished by reaction sintering at 1600°C for 2 h. Fly ash is also acceptable for incorporation into pottery and ceramic tile manufacturing. Fly ash-activated carbon is considered more cost-effective and applicable to the removal of odor and organic contamination at waste water treatment plants. For concrete applications, the use of fly ash is restricted to those in accordance with ASTM C 618 specifications. Class "F" ash with high carbon content may result in concrete with low air content, higher water-to-cementitious ratio, stale appearance, and variable consistency. Although specifications usually do not allow more than 25% fly ash for incorporation into concrete (e.g., Michigan Department of Transportation), up to 40% has been experimentally demonstrated as acceptable without jeopardizing quality, performance, and workability.

In addition to uses in asphalt and cement, fly ash can serve a number of other uses, including roles in the manufacturing of bricks, landfill liner, agriculture, and reclamation. When utilized as an ingredient in the production of bricks, up to 25% fly ash and 75% clay are used. Fly ash brick and/or tile plants are operational in California, North Dakota, Alberta, Canada, and Czechoslovakia. The plant in Alberta, Canada, is designed to produce 35 million bricks yearly. The plant in Czechoslovakia is designed to consume 100,000 tons of fly ash annually. Fly ash also shows promise for use as an agricultural fertilizer, revegetation additive, neutralizing diluent, and soil amendment to stressed areas. In West Virginia, about 9350 tons of fly ash were used over certain areas of a sanitary landfill due to its relative compaction density and high pozzolanic action, which increases with time. Although experimental, cenospheres are being studied and considered as a closed-pore insulation material for use on the space shuttle, for fire-proofing and insulating high-voltage electrical cables.

10.5 LOW RADIOACTIVE WASTE (LLW) AND INTERMEDIATE RADIOACTIVE WASTE (ILW)

The objectives for solidifying or conditioning low- and intermediate-level radioactive waste via bituminization is to reduce the risk of exposing personnel to radiation, contain the radionuclides within a durable product of low leachability, reduce the volume and weight suited for safe and final (geologic) disposal, and minimize transport and storage costs. The end product is containerized for the safe disposal of LLW and ILW with time in a geologic repository.

The first thorough study on the use of asphalt in the field of nuclear waste management was performed at the Oak Ridge National Laboratory, Oak Ridge, Tennessee, in 1958. It was concluded that asphaltic membranes appeared practical for the lining of earth storage pits for aqueous radiochemical waste, provided that the wastes (1) were neutralized and (2) were decayed sufficiently so that the self-heating temperatures did not exceed 65°C (150°F), and that the time for the asphalt to acquire a dose of 10^9 rad was more than 25 yr.

Since 1960, bitumen has increasingly been used for the incorporation of LLW and ILW, a process referred to as bituminization. The first investigation on the bituminization of radioactive wastes was carried out at the Research Centre for Nuclear Energy at Mol, Belgium, and at the Plutominum Production Centre at Macoule, France. Since the mid-1960s, facilities for the bituminization of radioactive waste have been developed in many countries, including the United States, Soviet Union, United Kingdom, Sweden, Denmark, Austria, Switzerland, Finland, Poland, Japan, and Czechoslovakia. The successful and safe bituminization of a wide variety of LLW and ILW generated at reactor sites and reprocessing plants has thus been demonstrated for more than 25 yr on an industrial scale using various pretreatment processes and incorporation techniques and conditions.

As of the late 1970s, several thousands of 200 l drums filled with bitumen waste products (BWPs) up to about 1 Ci/l have been produced in the industrial plants operating in Belgium, France, the Federal Republic of Germany, Finland, and Sweden. Bitumens were utilized as fixation material for certain types of radwaste due to their favorable chemical and physical properties. Such properties in comparison to other conditioning methods are preferred because bitumens produce durable, homogeneous products of low leachability, volume, and weight, and are well suited for safe final storage, within limits, independent of local geologic and hydrogeologic conditions.

Solid and initially liquid wastes of LLW with ILW can undergo bituminization via incorporation in or coating with bitumen. As with hydrocarbons and metals, bituminization is relatively insensitive to the type of waste being processed and can be operated either as a continuous or a batch-type process.

Bituminization of solid and initially liquid wastes can be accomplished by these different incorporation techniques.

- **Mixing with molten bitumen.** Mixing of the solid or liquid wastes with molten bitumen at temperatures of 140 to 230°C concurrent with evaporation of water and casting of the fluid mixture into containers or drums for storage after cooling to a solid product.
- **Mixing with emulsified bitumen.** Mixing of the solid or liquid wastes with emulsified bitumen at room temperature and subsequent

heating of the mixture obtained to evaporate water. The remaining mixture is then containerized, once it is cooled down, and stored.

- **Mixing with surface-active agents and bitumens.** Mixing of waste sludges, surface-active agents for emulsifying, and bitumen to initially obtain a preliminary coating with simultaneous partial removal of water. The final bitumen coating of the mineral salts and other solids is then removed. This is achieved by heating the mixture to about 130°C. The solids–bitumen mixture is then containerized, once it is cooled, and stored.

The high weight and volume percentage of the solid waste components in the final product are highly economical since the end product results in reduced costs in respect to embedding materials (bitumens), transport, handling, and storage. Acceptable compositions for long-term behavior are composed of about 50% in weight of LLW or ILW (reactor or reprocessing wastes) and about 50% in weight of a suitable type of bitumen.

Processing equipment primarily includes reaction vessels, evaporators, and mechanical mixer, in addition to condensers, off-gas cleanup systems and condensate cleanup systems (e.g., filters for the removal of bituminous oils). Heating and mixing equipment that has been used on an industrial scale includes pots with a heating mantle or interior heating elements and a mechanical stirrer for batch processes, two- and four-screws extruder evaporators, and/or wiped-film or thin-film evaporators. The extruder evaporator provides effective continuous mixing concurrent with evaporation of water and volatile components and extrusion of the resultant waste-bitumen mixture in product containers. Wiped-film evaporators also allow continuous mixing of the solid residue with bitumen, after evaporation of the water, along the heated evaporator walls, prior to being containerized. A schematic of the bituminization facility at the Riso research facility in Denmark is shown in Figure 10-1.

The majority of the bitumen-waste products (BWP) are composed of about 40 to 50% by weight bitumen with the remainder being the solid constituents originally present in the wet LWR wastes. These solid constituents are mainly spent ion exchange resins, filter sludges, and evaporator concentrates. Ion exchange resins are typically in bead form or powder resins, which are bead resins that have undergone grinding. The water associated with the ion exchange resins is partially or almost completely removed utilizing a dryer prior to mixing with liquefied bitumen, whereas water associated with the aqueous solution and slurries is removed during the incorporation process until a residual content of less than 1% by weight is achieved. The final product is thus a mixture of bitumen with the solid and nonvolatilized constituents, at the incorporation temperatures and time, of the treated wastes. Properties of typical LLW and ILW reprocessing waste-bitumen mixtures are presented in Table 10-2.

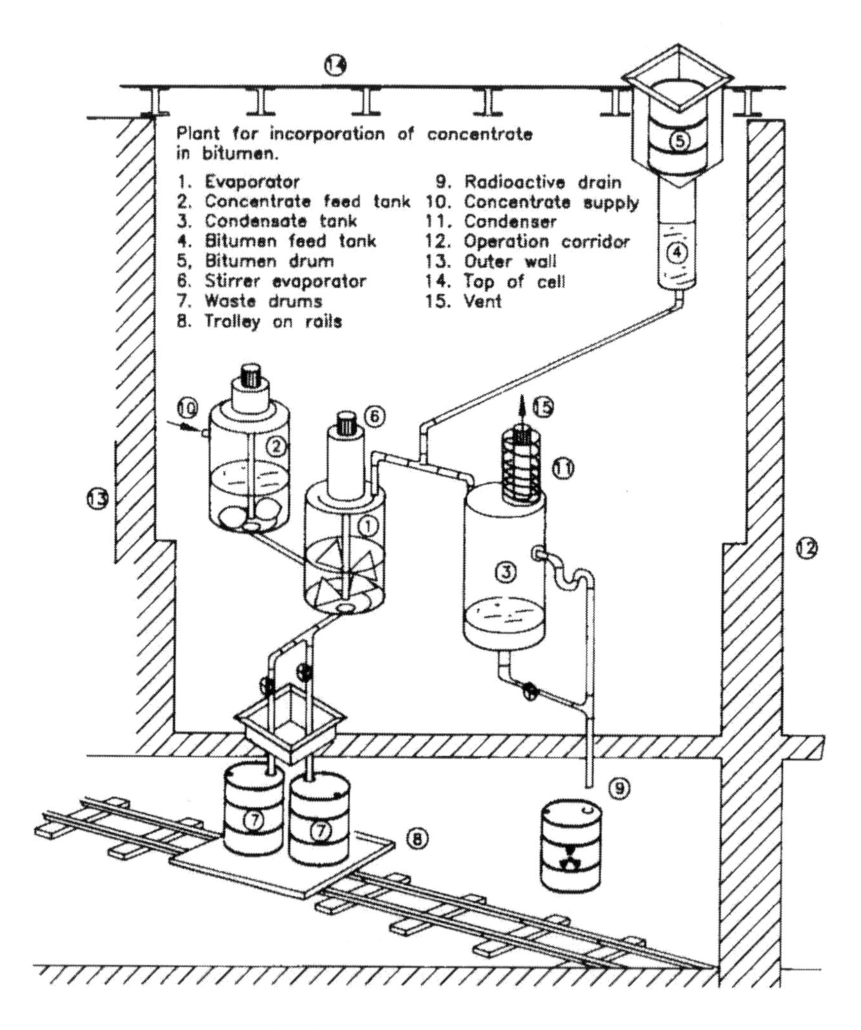

Figure 10-1. Schematic of the bituminization facility at the Riso research facility in Denmark.

Bitumens maintain numerous favorable physicochemical properties for their application as embedding materials for nuclear wastes. In general terms, bitumens are thermoplastic, readily adhesive, easily liquefied upon heating, highly waterproof, durable, resistant toward most acids, alkalis, and salts, nonpoisonous, highly resistant toward aging and climatological influences, and soluble in many organic solvents and light. These favorable properties and their long-term durability form the basis for their application for incorporation and containerization of nuclear waste, with a sufficient degree of reliability.

Table 10-2 Some Properties of Typical LLW and ILW

		ILW		LLW		
		Mechanical head-end	Chemical head-end			
Properties	Unit	55% Blown bitumen ~45% salts 0.3% H_2O	55% Blown bitumen ~45% salts 0.3% H_2O	BWR/PWR 50% bitumen 48% ion exchange ~0.3% H_2O	PWR 53% bitumen ~47% salts ~0.3% H_2O	BWR 50% bitumen ~35% salts ~15% fillers ~0.5% H_2O
Specific density	$9/cm_3$	1.36	1.35	1.18	1.40	1.35
Softening point	°C	~115	>110	90	90	>85
Flash point	°C	>290	>290	>300	>300	>300
Ignition point	°C	~380	>360	~400	~400	~425
Leachability[a]	$9 \cdot cm^{-2} \cdot d^{-1}$	$\sim 10^{-4}$	$\sim 10^{-5}$	3×10^{-5}	8×10^{-4}	1×10^{-3}
Specific activity	Ci/l	~0.8	~1	0.05	<0.01	<0.02
Heat generation	W/l	<0.005	<0.005	$<3 \times 10^{-4}$	$<5 \times 10^{-5}$	$<10^{-4}$
Total integrated dose	rad	$<2 \times 10^{8}$	$<2 \times 10^{8}$	$<10^{7}$	$<2 \times 10^{6}$	$<4 \times 10^{6}$
Spec. H_2-generation rate	$cm^3/Mrad \cdot gBWP$	<0.005	<0.005	<0.005	<0.005	<0.005

[a] In distilled H_2O; average after 1 yr.

Modified from Eschrich, H., 1980, Properties and Long-Term Behavior of Bitumen and Radioactive Waste — Bitumen Mixtures, Eurochemic, Mol, Belgium, 174 pp.

BIBLIOGRAPHY

Ciesielski, S. K. and Collins, R. J., Current Nationwide Status of the Use of Waste Materials in Hot Mix Asphalt Mixtures and Pavements, Eurochemic, Mol, Belgium, 174 pp.

Eschrich, H., 1980, Properties and Long-Term Behavior of Bitumen and Radioactive Waste — Bitumen Mixtures, Eurochemic, Mol, Belgium, 174 pp.

Fahnline, D. E. and Regan, R. W., Sr., 1995, Leaching of metals from beneficially used foundry residues into soils, in 50th Purdue Industrial Waste Conference Proceedings, Ann Arbor Press, Chelsea, MI, pp. 339–347.

Ham, R. K. et al., 1994, Evaluation of Foundry Wastes for Use in Highway Construction, Final Report to Wisconsin Department of National Resources and Transportation, Madison, WI.

Hustwit, C. C., 1995, Solidification, Foundry and Combustion Residues, 50th Purdue Industrial Waste Conference Proceedings, Ann Arbor Press, Chelsea, MI, pp. 329–337.

Iskander, I. K. and Selim, H. M., Eds., 1992, *Engineering Aspects of Metal — Waste Management*, CRC/Lewis, Boca Raton, FL, 231 pp.

Kandhal, P. S., 1993, Use of waste materials in hot mix asphalt — an overview, in *Use of Waste Materials in Hot-Mix Asphalt* (Edited by H. F. Waller), ASTM, STP, 1193, pp. 3–17.

Kramer, R. S., Hwang, J. Y., Huang, X., and Hozeska, T., 1994, Characterization of recyclable components in fly ash, in Extraction and Processing for the Treatment and Minimization of Wastes (Edited by J. Hager et al.), pp. 1075–1096.

Means, J. J., Heath, E. B., Marlux, K., and Solare, J., 1991, The feasibility of recycling spent hazardous sandblasting grit into asphalt concrete, in *Waste Materials in Construction*, Elsevier Science, Amsterdam, The Netherlands.

Torrey, S., Ed., 1978, Coal Ash Utilization — Fly Ash, Bottom Ash and Slag, Noyes Data Corporation, Park Ridge, NJ, 370 pp.

United States Environmental Protection Agency, 1991, Treatment of Lead-Contaminated Soils, U.S. EPA Report No. EPA 540/2-91/009, April, 1991, 10 pp.

Westsik, J. H., Jr., 1984, Characterization of Cement and Bitumen Waste Forms Containing Simulated Low-Level Waste Incinerator Ash, NUREGICR-3798, PNL-5153, Pacific Northwest Laboratory, Richland, WA.

11 ASPHALT UTILIZATION AND APPLICATION

11.1 INTRODUCTION

The incorporation of contaminated soil and other materials into asphalt can at best be described as user-friendly. There currently exists a multitude of uses, as discussed in previous chapters, for HMA and CMA incorporating affected soil. Presented in this chapter is further discussion for the on-site or off-site use of the end asphaltic product as pavement, liners, berms, and hydraulic applications.

Application considerations, such as seal coating and curing, and quality control and assurance including production consistency, temperatures, weight, cost, and certificate of compliance are also discussed.

11.2 UTILIZATION

11.2.1 Use as Pavement

In 1991, approximately 2,060,000 metric tons (2,270,000 tons), or 9300 lane-km (5800 lane miles), of CMA utilizing reclaimed asphalt were processed in the United States. CMA is routinely used in California, Kansas, New Mexico, and Oregon. CMA is also frequently used on medium- to low-traffic-volume roadways. New Mexico, for example, uses 75 to 125 mm (3 to 5 in.) of HMA on top of the CMA layer to accommodate truck traffic. Excluding reclaimed asphalt, no definitive numbers are available as to the amount of CMA incorporating contaminated soil that is produced. However, this number has been steadily increasing, reflecting the increased number of recycling companies established within the last several years.

The most traditional use of produced CMA and HMA asphaltic products incorporating contaminated soil is as pavement, which use can best be described as user-friendly. Visualize a typical multilane high-traffic-volume freeway and the load-bearing capability and durability that must be designed into the asphalt product used in its construction. Now visualize the typical bicycle path winding its way through our urban areas. The point is that both the freeway and the bicycle path are asphalt pavements, but their end uses are drastically different. CMA addresses such variability in the end product best; CMA is also certainly nothing new. There are very few if any state and county road departments that do not use variations of CMA. The end use of asphalt dictates its specifications, or, better stated, if the asphalt mix will perform its required function, from freeway to bicycle path, it is within specifications. In fact, ASTM procedure for cold-mix asphalt design includes a section that states that the mix must fulfill the requirements of its intended application. Recalling the phrase user-friendly, it becomes apparent that the function of the end product will determine the asphalt mix design.

Application of processed asphalt is typically performed in 3-in. lifts, with an asphalt emulsion tack coat applied to the native surface and between lifts. Mixes for lay-down application can be open-graded, dense-graded, or sand mixes.

Open-graded mixes are characterized by high permeability, high void content, and flexibility. Thus, open-graded mixes have been used for years for the rapid removal of surface water, for drainage layers, and to reduce hydroplaning. A variety of aggregate gradations have been used for open-graded mixes, with mixed results in regard to performance. Asphalt emulsions most commonly utilized are MS-2, MS-2h, HFMS-2, HFMS-2h, HFMS-2s, CMS-2, or CMS-2h.

Dense-graded mixes include gradation from the maximum size down to and including material passing the 75 μm (No. 200) sieve size fractions. Dense-graded mixes are characterized by low permeability and low void content, and high strength and durability suitable for high load-bearing surfaces and is comparable to HMA concrete. Asphalt emulsions typically used for such mixes include MS, CMS, SS, CSS, and HFMS.

Sand mixes are used for either base or surface construction. Aggregate gradation requirements are similar to those for open- and dense-graded mixes. Asphalt emulsion can vary from between 6 and 15%. Emulsion types include SS-1, SS-1h, CSS-1, CSS-1h, and HFMS-2s.

An example of such use is pavement utilized for a heavy equipment yard that was constructed from asphalt made with affected soil recovered from leaking underground tanks. By producing parking lot pavement for on-site use, the generator eliminated the inherent liability of disposing of its contaminated soil in a dump site. Approximately $80.00 per ton of disposal taxes were saved as the materials were recycled and not disposed of. The pavement produced not only kept the project's pricing below any other option but created a paved parking lot of extremely low permeability to prevent further adverse subsurface impact. The mix design was not the same as that required to construct a freeway, but then a freeway was not the intended use. The intended use was for low traffic volume but required extremely high load-bearing strength. Another project used affected soil from an oil tank spill for load-bearing pavement at an oil refinery (Figures 11-1 and 11-2). The affected soil

Figure 11-1. CMA incorporating petroleum hydrocarbon-contaminated soil derived on-site being applied to pregraded area at a refinery.

was not disposed of as hazardous waste but rather was recovered and used in a CMA pavement with the mix design consistent with the end use.

The minimum Marshall stability required for paving mixtures, for example, is 2224. Mix designs used for actual applications range from a 95%

Figure 11-2. Load-bearing operation road made of petroleum hydrocarbon-contaminated soil derived on-site at a refinery.

contaminated soil (native silt, sand, and gravel contaminated with diesel fuel to 32,000 ppm total petroleum hydrocarbons) with a 5% emulsion. CMA has been successfully used on a variety of projects, ranging from road base and road pavement to containment dikes and drain channels. With CMA, the procedure is to determine the requirements and then develop the mix design to fit the use. As the equipment used to produce a CMA asphaltic product is portable and certainly not complex, isolated areas with limited access are most suitable for portable CMA units (Figure 11-3). Field test batches of 20 tons or more can be used rather than bench-scale tests. In this manner, the actual field mix is tested rather than a small hand-mixed batch. The bearing ratios for environmentally processed asphalt for Class II base is presented in Figure 11-1.

Where stringent specifications must be followed for pavement design, tests have been developed for both HMA and CMA. These tests, and their respective properties measured, are presented in Table 11-1.

11.2.2 Use as a Liner

The imminent closure of many of the nation's Class III and municipal landfills creates a potential use for hundreds of thousands of tons of contaminated soils incorporated into asphalt for use as a landfill liner or cap. The cost effectiveness of this viable and creative method of capping landfills is very attractive to financially strained municipalities. Prior to the advent of envrionmentally processed asphalt for use as a liner or a cap, clay was the specified material. In

Figure 11-3. Portable CMA unit situated along interior perimeter of a refinery reservoir engaged in processing of petroleum hydrocarbon-contaminated soil for on-site use as pavement.

Table 11-1 Summary of Tests for Asphaltic Paving Materials

Test	Property	Test method
Standard specification for fine aggregate for bituminous paving mixtures	Soundness Uniformity Size Grading Classification	ASTM D-1073-88
Standard classification for sizes of aggregate for road and bridge construction	Classification Size Grading	ASTM D-448-86
Standard specification for hot-mixed, hot-laid bituminous paving mixtures	Mix design Recycled asphalt mix design Recycled aggregate mix design Equipment specification	ASTM D-3515-83
Standard specification for cold-mixed, cold-laid bituminous paving mixtures	Mix design Equipment specification	ASTM D-4215-87 (Reapproved 1992)

addition to environmental concerns associated with mining vast quantities of clay for these uses or the problems in obtaining readily available deposits of clay of low hydraulic conductivity, there were no cost-recovery options for landfill operations or closure. By using environmentally processed asphalt, municipalities and landfill owners can charge attractive fees to accept affected soil. In most cases, this acceptance fee pays for the cost of on-site processing of the affected soil into the asphalt end product. The effectiveness of the cost recovery is obvious as the capping materials production process becomes a profit center. By using on-site material, not only is the cost of obtaining the clay avoided, but transportation costs also are eliminated. In essence, the capping process for landfill closure is more affordable, makes use of a product far superior to the traditional clay method, and reduces a broad spectrum of environmental concerns by keeping affected soil out of landfills as a waste. Instead it places affected soil as environmentally sound end products, such as caps or liners.

There exists a wide range of liner types and combinations of liners that can be used for landfills, impoundments, ponds, etc. Liner system performance is measured by the liner's ability to minimize seepage. Seepage minimization is dependent upon the intrinsic properties of the liner material and upon loading conditions, quality of construction, and characteristics of the underlying natural or waste geologic materials. Selecting the most appropriate liner material will require consideration of climatic conditions, waste properties, water resource values, hydrogeologic characteristics, and proposed waste management practices.

Studies of asphalt, clay, and other membrane liners subjected to a variety of aging tests in exposure columns at various temperatures, pH conditions, oxygen concentrations, and hydrostatic pressures have been discussed. The conclusions were that the asphalt liners and membranes were extremely stable chemically and physically. An aging period equivalent to 7 yr produced penetration of reaction products to only 0.5 mm (0.5% of the 10-cm liner thickness). The results showed that if the asphalt content of the liner exceeded about 6%, these liners would perform adequately under impoundment conditions for over 1000 yr, conditions that are similar to those expected for CMA. Catalytically blown asphalt was considered the best liner material and was selected for long-term field testing. Field tests of catalytically blown asphalt over a 2-yr period showed superior performance of the asphalt liners compared to clay liners. This will be especially true for the petroleum constituents in CMA liners; overall, asphalt is a much better liner material for this application than clay.

Studies of asphalt, clay, and other membrane liners subjected to a variety of aging tests in exposure columns under various temperature and pH conditions, oxygen concentrations, and hydrostatic pressures have been previously addressed. Asphalt liners and membranes were found to be extremely stable both chemically and physically. An aging period equivalent to 7 yr produced penetration of reaction products to only 0.5 mm (0.5% of the 10-cm liner thickness). The results showed that if the asphalt content of the liner exceeded 6%, these liners would perform adequately under impoundment conditions for more than 1000 yr, conditions which are similar to those expected for CMA.

Catalytically blown asphalt was considered the best liner material and was selected for long-term field testing. Field tests of the catalytically blown asphalt over a 2-yr period showed superior performance of asphalt liners over clay liners. This would be especially true for the petroleum constituents in CMA, with asphalt serving as a much preferred liner material for this application than clay.

Permeability tests performed on a variety of liners after being subjected to aging tests also showed favorable results. Accelerated aging tests of an asphalt liner at 20°C under oxygen partial pressures of 0.21, 1, and 1.7 atm, with continuous exposure to an acidic leachate, have also been performed. The composition of the leachate was

CaSO	500 g/l
CaCO	18.1 g/l
MgSO	8.6 g/l
NaSO	7.4 g/l
NaCL	7.4 g/l
FeO	2.4 g/l
NaCO	2.3 g/l
AlO	1.2 g/l

Table 11-2 Anticipated Field Linear Permeabilities

Liner	Avg. final permeability, *K* (cm/sec)	Assumed field thickness, *L* (cm)	Effectiveness factor, *K/L* (sec^{-1})
CMA	7×10^{-8}	10	7×10^{-9}
Hypalon	2×10^{-10}	0.12	2×10^{-9}
Asphalt rubber membrane	4×10^{-6}	0.8	5×10^{-6}
Catalytic air-blown membrane	7×10^{-9}	0.9	8×10^{-9}
Sodium bentonite	1×10^{-7}	10	8×10^{-8}
Saline seal 100	8×10^{-6}	10	6×10^{-7}
GSR-60	6×10^{-6}	10	6×10^{-7}
Soil (as a liner)	1×10^{-5}	10	1×10^{-6}

Solution pHs of 2.5, 2.0, and 1.5 were designated as normal, intermediate, and highly accelerated conditions. Acidity levels were shown to have an unmeasurable effect on asphalt aging. Permeability was used as a means to measure the immediate effectiveness of the asphalt liner. Liner permeabilities under these exposure conditions appear to be relatively unaffected, as presented in Table 11-2. The relative costs of emplacement per liner type is presented in Table 11-3.

HMA, as previously mentioned, must adhere to very restrictive, stringent specifications. HMA has, however, been used as both an impermeable asphaltic cap or liner while also serving as a surface for heavy duty traffic and load-bearing pavement. Such usage occurred at an arsenic-based pesticide manufacturing facility in the Bay Area of California. In-place soils were mixed with

Table 11-3 Estimated Costs of Candidate Linear Materials

Liner	Installation rate	Installed costs
Asphalt concrete	Two 5-cm lifts with cationic asphalt emulsion tack coat (3 l/m^3)	\$9.20/m^2
Chlorosulfonated polyethylene	0.76 mm (30 mil)	\$5.90/m^2
Asphalt-rubber	5.4 l/m^2	\$2.40/m^2
Catalytic airblown asphalt	9 l/m^2	\$2.60/m^2
Saline Seal 100	20 kg/m^2	\$4.70/m^2
GSR-60	20 kg/m^2	\$2.70/m^2
Sodium bentonite	20 kg/m^2	\$2.30/m^2

Note: Assume that 10^3 2-cm (4-in.) layer of clay and soil is installed with soil at \$5.20/m^3.

silicates and Portland cement to a depth of 8 m (26 ft), reducing the leachability of the arsenic and raising the pH of the cement. The mixture eventually behaved much like cement-treated base. Due to the presence of a shallow water table, the liner had to both serve as an impermeable liner, preventing water from percolating through the soil column to the groundwater, and structurally sustain the weight of 18-wheel truck traffic and construction storage. A three-layer HMA pavement was designed for the cap — a 75-mm (3-in.) thick hydraulic mix for the lower layer, a 75-mm (3-in.) open-graded asphalt drainage middle layer, and a conventional 75-mm (3-in.) thick dense grade of Caltrans mix used for heavy duty highways and industrial pavements. The lower layer mix design was designed with high asphalt content and low air voids, thus providing an impervious layer to water while maintaining sufficient strength without deformation under heavy traffic loads. The middle layer served as a drainage path for any potential water entering the pavement, which would ultimately drain toward the bay. A schematic of the liner/pavement is shown in Figure 11-4.

11.2.3 Use as a Berm

Berms are used to contain runoff or unanticipated breach of fluids from a holding tank and are a common fixture at tank farms and terminals, refineries, and other industrial complexes. Although the dimensions of berms vary with the estimated amounts of fluids these structures are expected to retain, they usually must be constructed to contain storm water runoff for a 100-yr average 24-h rainfall event or to contain the fluid contents of a breach tank(s) which the berm encloses. At some facilities such as refineries, a particular site may contain miles of berms originally constructed when the facility was built. Many

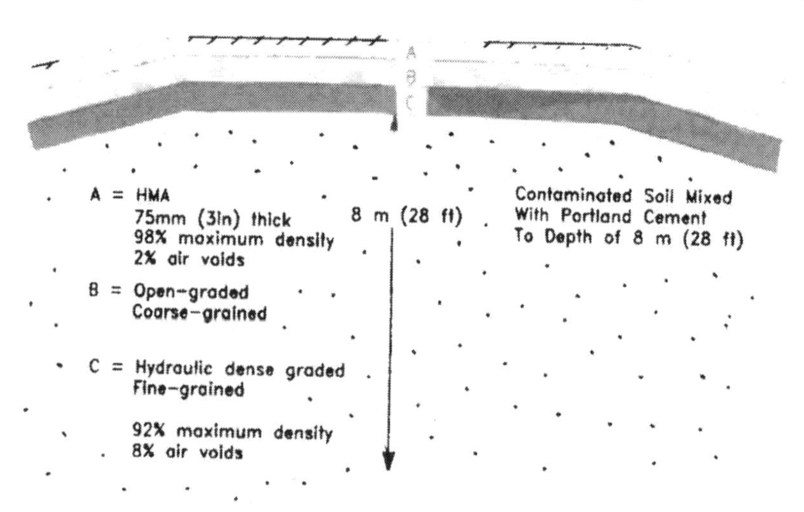

Figure 11-4. Schematic showing liner/pavement design for HMA incorporating contaminated soil.

Figure 11-5. Degraded asphalt berm showing unlined tankage area and broken-up, rodent-ridden berm at an active refinery site.

of these structures need replacing and/or maintenance, due to erosion, degradation, and rodents (Figure 11-5).

The use of contaminated soil as an ingredient in asphalt serves this purpose quite well. Most facilities of this nature have large quantities of contaminated soil. These facilities also have need for asphalt for capital improvements such as berms and service roads (Figures 11-6 and 11-7). An added benefit is the containment of runoff water, which can also be utilized for dust abatement, landscape irrigation, and compaction associated with geotechnical projects.

11.2.4 Hydraulic Applications

HMA has been used in hydraulic applications, such as domestic water reservoirs, fish rearing ponds, and canal liners, for more than 50 years. The Metropolitan Water District of southern California (MWDSC) has employed HMA to line treated and untreated water reservoirs for more than 4 decades. The East Bay Municipal Utility District in Oakland, California, has been using HMA to line domestic water supply reservoirs since the 1950s. In Oregon and Washington, fish and wildlife agencies have used HMA to line fish rearing ponds since 1987. Such ponds are typically 0.5 acre in area, where chinook salmon and other fry are reared for about 18 months before the fish are released into a more natural setting.

A 50-acre evaporation pond composed of a HMA liner was constructed by Southern California Edison Company (SCEC) at its Mohave Generating Station. Other liners have also been constructed by SCEC for sweet water and coal evaporation ponds.

Figure 11-6. Initial application of CMA for a berm utilizing petroleum hydrocarbon-contaminated soil at an above-ground tank area at an active refinery site.

Figure 11-7. Compaction of CMA for a berm utilizing petroleum hydrocarbon-contaminated soil at an above-ground storage tank area at an active refinery site.

Dense-graded HMA is typically used for hydraulic applications, with a relatively higher asphalt content in comparison to mixes used for road pavement. Mixes that easily compact with a lower air void content are preferred. In some cases, a built-in subdrain system is designed, consisting of two dense-graded layers, with an intervening open-graded layer. The open-graded layer serves as a drainage and leak detection layer between the two impermeable, dense-graded asphalt layers.

The use of CMA for hydraulic applications is not common. CMA incorporating contaminated soil has, however, been successfully used for berms and above-ground tank farm containment areas, as previously discussed. Typical gradations for asphalt mixtures in hydraulic applications are presented in Table 11-4.

11.3 APPLICATION

CMA mix designs can easily be worked into open-graded, dense-graded, and sand mixes. Open-graded mixtures have been used routinely for sub-bases and surfaces for years. Such mixtures exhibit flexibility and high void contents, making them highly resistant to fatigue and reflection cracking. When CMA mixtures are used as base material, a positive moisture seal is used within and under the layer to prevent water from migrating to subgrade materials, which

Table 11-4 Typical Gradations for Asphalt Mixtures in Hydraulic Applications

Sieve size		Mix type[a]	
(mm/μm)	(in.)	Dense-graded (low voids)	Open-graded (high voids)
25.0 mm	1	—	100
19.0 mm	3/4	—	93–100
12.5 mm	1/2	100	—
9.5 mm	3/8	95–100	35–65
4.75 mm	No. 4	70–84	5–25
2.36 mm	No. 8	52–69	2–15
1.18 mm	No. 16	38–56	0–7
600 μm	No. 30	27–44	0–3
300 μm	No. 50	19–33	—
150 μm	No. 100	13–24	—
75 μm	No. 200	8–15	—
Asphalt cement, percent by weight of total mix		6.5–9.5	1.0–4.0

[a] Percent passing.

Modified from Asphalt Institute (1990).

are adversely susceptible to water. When used as a surface layer, open-graded mixes permit the rapid removal of surface water, due to their relatively higher permeability.

Dense-graded mixtures can accommodate the full range of base and surface pavement types, depending on aggregate type. These mixes typically utilize all size fractions, including material passing the 75 mm (No. 200) sieve.

Sand mixes are similar to dense-graded mixes, except for aggregate gradation.

11.3.1 Seal Coating

Seal coating, while not required from a technical perspective, is commonly applied to the laid-down surface for aesthetic purposes. When a water-based emulsion is used in the production of CMA, and all bituminous compounds in fact, the product produced is not black but brown. Lamp black is usually added to HMA to provide a uniform black color. With CMA, the application of a seal coat provides the more aesthetic black coloring. Seal coating also inhibits exposure to UV rays and smog. A summary of seal coats and surface treatments is presented in Table 11-5.

11.3.2 Curing

A sufficient time for curing should be allowed prior to loading. Typically 7 to 14 d is sufficient, although in most cases 2 to 3 d is adequate for normal loads, depending on weather conditions.

11.4 QUALITY CONTROL AND ASSURANCE

Quality control and assurance can be discussed generally in regard to production consistency, temperatures, weight, cost, and certification of compliance. These factors are discussed below.

11.4.1 Production Consistency

During on-site production of CMA, a minimum of three representative asphalt samples should be retrieved daily, composited, and hand-compacted using conventional Marshall Stability testing methodology. Following time allowed for curing, approximately 3 d, the samples should then be submitted to a certified laboratory for chemical testing for the constituents of concern. It is also a good precaution to separate and label each day's production to facilitate ongoing use of the final product.

11.4.2 Temperature

When transporting to the site via truck or tank car, the temperature of the emulsion cannot be raised above 160°F nor cooled to a temperature less than

Table 11-5 Emulsified Asphalt Seal Coats and Surface Treatments

Type of construction	Description	Uses	Typical emulsified asphalts	Construction hints
Sand seal	Restores uniform cover; enriches dry, weathered pavements; reduces raveling.	In city street work; improves street sweeping; traffic line visibility.	CRS-1, CRS-2, RS-1, RS-2, MS-1, HFMS-1	Spray-applied with sand cover. Roll with pneumatic roller. Avoid excess binder.
Chip seal	Single most important low-cost maintenance method.	Produces an all-weather surface; renews weathered pavements; improves skid resistance; lane demarcation; seals pavement.	CRS-2 or RS-2	Spray-applied Many types of textures available. Key to success: coordinate construction, use hard, bulky grained, clean aggregate, and have properly calibrated spray equipment.
Double seal	Two applications of binder and aggregate. The second chip-application uses a smaller-sized stone than the first.	Provides some leveling; available in a number of textures; durable.	CRS-2 or RS-2	Similar to chip seal.
Triple seal	Three applications of binder and three sizes of chips are applied to provide up to a 20 mm (¾ in.) thick, flexible pavement.	Levels as well as providng a seal; tough-wearing surface.	CRS-2 or RS-2	Spray-applied in three lifts.

Table 11-5 Emulsified Asphalt Seal Coats and Surface Treatments *(continued)*

Type of construction	Description	Uses	Typical emulsified asphalts	Construction hints
Slurry seal	Used in airport and city street maintenance where loose aggregate cannot be tolerated.	Seals, fills minor depressions, provides an easy-to-sweep surface. The liquid slurry is machine-applied with a sled-type box containing a rubber-edged strike-off blade.	CSS-1, CSS-1h, SS-1, SS-1h, or QS[a] grades	Pretest the aggregate and emulsion mix to achieve desired workability, setting rate, and durability. Calibrate equipment prior to starting the project.
Cape seal	Combines a single chip seal with a slurry seal.	Provides the rough, knobby surface of a chip seal to reduce hydroplaning yet has a tough sand matrix for durability. Test track data indicate better studded tire damage resistance than a chip seal.	CS-1, CSS-1h, SS-1, SS-1h, RS-2, CRS-2, and QS[a] grades	Apply an aggregate single chip seal. Broom and apply slurry seal. Have the strike-off ride on the rock surface to form the matrix. Avoid excess slurry as this destroys the knobby stone texture desired.

[a] The quick-set grades of emulsion (QS) have been developed for slurry seals. While not yet standardized, their use is rapidly increasing, as the unique quick-setting property solves one of the major problems associated with the use of slurry seals.

Modified from The Asphalt Institute (1990).

Table 11-6 Typical CMA Temperatures for Construction Purposes

Emulsion type	Emulsion grade	Temperature: Mixing facility	Temperature: Windrow prior to mixing
Emulsified asphalt	Anionic MS-1, MS-2, MS-2h, SS-1, SS-1h, HFMS-2, HFMS-2h, HFMS-2s	10–70°C (50–160°F)	20–70°C (70–160°F)
	Cationic CMS-2, CMS-2h, CSS-1, CSS-1h	10–70°C (50–160°F)	20–70°C (70–160°F)
Cutback asphalt	MC, SC asphalts		
	70	—	20°C+ (65°F+)[4]
	250	55–80°C (135–175°F)[3]	40°C+ (105°F+)[4]
	800	75–100°C (165–210°F)[3]	55°C+ (135°F+)[4]
	3000	80-115°C (175–240°F)[3]	—

Modified from Asphalt Institute (1990).

40°F. Application temperatures for CMA vary, depending upon whether the asphalt is metered through a mixing facility or windrowed, and depending upon the type of emulsion being used. Typical CMA asphalt temperatures for construction use are presented in Table 11-6.

11.4.3 Weight

Asphaltic emulsions are measured in tons. When scales are not available, weight can be determined from volumetric measurements, provided that the tanks are calibrated and a proper measuring stick and calibration card or a proper vehicle tank meter are available.

11.4.4 Cost

Costs for asphaltic emulsions are routinely measured per ton. When water is added, the cost for emulsion is determined prior to the addition of the water. Typical costs range from $125 to $150 per ton, although cost varies depending on emulsion type, quantity being used, and availability. Payment for asphaltic emulsions is assessed by determining the volume that the emulsion would occupy at 60°F, before converting the volumetric measurement to weight.

11.4.5 Certification of Compliance

A certificate of compliance is a form that should accompany each shipment of asphaltic emulsion received. Information included on such forms includes shipment number, type of emulsion, refinery, co-signee, designation, quantity, purchase order or contract number, and date of shipment. In addition, the certificate needs also to state that the emulsion complies with certain specifications as required for the intended end use, as appropriate. Should a certificate of compliance not be provided, appropriate tests need to be performed before the emulsion is used.

BIBLIOGRAPHY

Asphalt Institute, 1990, *Asphalt Cold Mix Manual*, Manual Series No. 14 (MS-14), 3rd ed., 52 pp.

Asphalt Institute, 1991, Asphalt Use in Water Environments, Information Series No. 186 (IS-186), December, 1991.

Dickson, G., 1996, Turning hydrocarbon contaminated soils into an asset, *Environ. Technol.*, July/August, 1996, pp. 68–70.

Hinkle, D. R., 1976, Impermeable asphalt concrete pond liner, *Civil Eng. ASCE*, August, 1976, pp. 56–59.

Hutchison, I. P. G. and Ellison, R. D., Eds., 1992, *Mine Waste Management*, CRC/Lewis, Boca Raton, FL, 654 pp.

Smith, R., 1995, Hot mix takes on double duty job, *Asphalt Contractor*, December, 1995, pp. 56–59.

Testa, S. M., Patton, D. L., and Conca, J. L., 1992, Fixation of petroleum contaminated soils via cold-mix asphalt for use as a liner, in Proceedings of the HMC South Conference, Hazardous Materials Control Research Institute, pp. 30–33.

United States Environmental Protection Agency, 1976, Gas and Leachate from Landfills — Formation, Collection and Treatment, U.S. EPA Report No. EPA 600/9-76-004, March, 1976, 158 pp.

12 SELECTED CMA CASE HISTORIES

12.1 INTRODUCTION

The incorporation of various contaminated soils as an ingredient to produce a variety of asphaltic end products shows the most promise in the reuse and recycling of both small and large quantities of contaminated soils. As such, a selection of case histories is provided in this chapter, illustrating the versatility of CMA technology. The cases present a variety of contaminant types, including petroleum hydrocarbons, polynuclear aromatic hydrocarbons (PNAs), metals (notably lead), and combined lead and petroleum hydrocarbons, respectively. The specific sites discussed include a former UST site, a gas manufacturing plant, a geothermal plant, a brake-shoe manufacturing plant, and an auto wrecking yard. The contaminants illustrated are fuel petroleum hydrocarbons, PNAs and MAHs, lead-mine scale, metals, and combined lead and fuel petroleum hydrocarbons, respectively.

Most reuse and recycling projects, as exemplified by these case studies, follow a similar sequence of activities. Pilot studies, as discussed in Chapter 6, are not always performed, depending upon the recycler's familiarity with the site, the contaminated soil, the contaminant type being incorporated, and intended end use. Excluding conduct of a pilot test, the typical sequence of tasks to be performed is outlined in Table 12-1.

Presented as part of each case history is discussion of site background, chemical engineering characteristics of preprocessed soil, mix design formulation, and chemical and engineering characteristics of the produced asphaltic product. Cost analysis information is also presented, as appropriate. A data summary for the case studies discussed and others is presented in Table 12-2. General discussion of cost consideration is also presented.

12.2 PETROLEUM HYDROCARBON-CONTAMINATED SOIL FROM MUNICIPALITY MAINTENANCE FACILITY (UST)

A city's yard serves as a vehicle maintenance shop and yard. These USTs and two fuel dispenser pumps were formerly located on the property. Upon

Table 12-1 Typical Task Summary for Reuse and Recycling of Contaminated Soil via CMA

Task no.	Description
1.0	Site reconnaissance
2.0	Acceptance criteria
2.1	Representative sample retrieval
2.2	Chemical screening
3.0	Cost-benefit analysis
4.0	Regulatory compliance analysis
5.0	Engineering testing of preprocessed soil
6.0	Chemical testing of preprocessed soil
7.0	Mix design formulation
8.0	Processing
9.0	Engineering testing of produced product
10.0	Chemical testing of produced product
11.0	Application
12.0	Confirmation of regulatory compliance

removal of the USTs and appurtenant structures, about 5689 yd^3 of clayey silt and clay were selectively excavated, of which about 4266 yd^3 were contaminated by diesel fuel, gasoline, and a tar-like substance. Total petroleum concentrations ranged up to 120,000 mg/kg. Total recoverable petroleum hydrocarbons ranged up to 4000 mg/kg. BTEX concentrations ranged up to 22, 1900, 70, and 7200 μg/kg, respectively. Volatile organic compounds concentrations ranged up to 486 ppm.

The mix design formulated consisted of 45% affected soil, 49% aggregate, and 6% water-based emulsion. Produced asphaltic samples retrieved were tested for Marshall stability, Marshall flow rate, and density. Marshall stability results ranged from 2600 to 2800 lbs., averaging 2667 lbs. Flow rates ranged from 0.25 to 0.33, averaging 0.29, consistent with the Asphalt Institute's specification of 0.20 to 0.40. Density as a measure of compacted mass (i.e., permeability potential) was relatively high, ranging up to 130.0 lb/ft^3 or 1.83 ton/yd^3. This is on the high end of the Type A open-graded hot-mix specification of 1.5 to 140 lb/ft^3, which equates to a permeability of 10^{-9} cm/sec. Such results exceeded that required for load-bearing asphalt road section base and sub-base materials, asphaltic stabilized base materials, pond/impoundment asphaltic liner material, pavement, or a variety of other usages.

Chemical testing for total petroleum hydrocarbons as diesel on the produced product was reported as nondetect. Bioassay results also showed favorable results, less than 40% fathead minnow dead in a 750 μg/l concentration.

Approximately 9600 tons of CMA were produced. Of this total, about 1200 tons were used on-site as pavement and designed fill, with the remainder

Table 12-2 Summary of Pertinent Performance Data for Reuse and Recycling via CMA at Various Sites

Site description	Contaminant type	Soil type	Soil volume (tons)	Analytical results: Parameter	Analytical results: Preprocessed soil	Analytical results: Processed CMA	Mix design soil/aggregate/emulsion/admis (% in total volume)	Emulsion type	CMA vol. (tons)	Engineering results: Preprocessed soil		Engineering results: Processed DMA: Marshall stability	Engineering results: Processed DMA: Flow rate	Usage
Municipality maintenance	Diesel	Silty clay (CL)	4266	TPH	120,000 mg/kg	ND	60/32/8	SS1h A48	5972	LAR	Grade A	2800	0.25	Street pavement
Yard (USTs)				TRPH	4000 mg/kg	ND				Moisture content	10.6–14.2%			
				Benzene	22 µg/kg	ND								
				Toluene	1400 µg/kg	ND				Unit weight	88–104.2 lb/cf			
				Ethylbenzene	70 µg/kg	ND				Dry unit weight	78.6–94.2 lb/cf			
				Xylenes	7200 µg/kg	NT								
				BioAssay	NA					Sand equivalent	25–43			
Former gas mfg. plant site	PNAs	Silty sand (SM)	Pilot test	TPH	1350 mg/kg	ND	60/33/7							
				TRPH	5120 mg/kg	ND	50/43/7	SS1h	NA	NA	NA	NA	NA	NA
				PNAs	260 mg/kg	ND	40/53/7							
				BioAssay	NA	NT	50/42/8							
Former geothermal plant	Lead mine scale	Silty clay (CL)	12,800 (pilot test)	Pb TCLP	7.59	ND	20/72/8/10	SS1h	NA	NA	NA	NA	NA	NA
				BioAssay	NA	NT	30/62/8/10							
							40/52/8/10							
							40/52/8							
							86/0/14							
Brake shoe mfg. plant	Lead zinc	Silty sand (SM)		Pb STLC	438 mg/l	60/32/8	SS1h	NA	NA	NA	NA	NA	NA	NA
				Zn STLC	336 mg/l	7.62 mg/l								
				Cd STLC	1.42 mg/l	ND								
				BioAssay	NA	NT								

Table 12-2 Summary of Pertinent Performance Data for Reuse and Recycling via CMA at Various Sites *(continued)*

				Analytical results						Engineering results				
												Processed DMA		
Site description	Contaminant type	Soil type	Soil volume (tons)	Parameter	Preprocessed soil	Processed CMA	Mix design soil/aggregate/emulsion/admix (% in total volume)	Emulsion type	CMA vol. (tons)		Preprocessed soil	Marshall stability	Flow rate	Usage
Auto wrecking site	Lead petroleum hydrocarbons	Silty sand (SM)	4100	TRPH	13,00 mg/kg	ND	60/32/7/1	A48	6539	NA	NA	3750	NA	Pavement
				Pb STLC	22 mg/l	ND								
				Pb TCLP	8.3 mg/l	ND								
				BioAssay	NA	NT								
Refinery	Petroleum hydrocarbons PNAs	Silty sand (SM)	5000	TPH	>1000 mg/kg	ND	60/32/8	A48	7000	NA	NA	3400	NA	Equipt. storage area
				TRPH	>1000 mg/kg	ND	60/32/8	SS1h	11,468					
		Sandy clay (SC)	8192	BioAssay	NA	NT								
Former gas mfg. plant	Petroleum hydrocarbons	Silty sand (SM)	3406	PNAs	91,000 mg/kg	ND	60/32/8	SS1h	4300	NA	NA	NA	NA	Secondary road
				BioAssay	NA	NT								

Note: ND = Nondetect; NT = nontoxic; NA = not applicable; cf = cubic feet.

being stockpiled and subsequently used for the city's roads and pavement for alleyways, as needed, over the 6-month period following processing.

12.3 POLYNUCLEAR AROMATIC HYDROCARBON-CONTAMINATED SOIL (PNAS) FROM FORMER GAS MANUFACTURING SITE

A former gas manufacturing plant operated at a site in Venice, California, since 1905. After 1912, availability of natural gas was poor, and the operation was significantly cut back. By 1915, it appeared that the site was no longer used for gas manufacturing. From 1916 to about 1950, the site was used for natural gas storage, system distribution, and administrative purposes. Specific functions included natural gas compression, meter painting and repair, and fleet vehicle maintenance, refueling, and storage.

Gas manufactured at the site used the oil-gas process, with no record of coal as feedstock. This process was based on the gasification of the oils by passing oil and steam through a chamber of heated checkerbricks. The process was cyclical, with alternate heating and gas manufacturing. The components most likely used included a boiler for steam supply, gas generator, superheater, washbox, and scrubber. The major by-products from the oil-gas process were lampblack, tar, and light oil. Small amounts of ammonia, cyanide, tar bases, and tar acids (i.e., phenols, creosols, etc.) were also produced.

Chemical testing of representative soil samples showed the presence of seven PNAs at concentrations such that the soils would be characterized as a hazardous waste. These constituents included anthracene, benzo(b) fluoranthene, benzo(a) pyrene, chrysene, fluoranthene, indeno (1,2,3,-cd) pyrene, and naphthalene. PNA concentrations of concern ranged up to 260 mg/kg, as illustrated in Figure 12-1. Total petroleum hydrocarbons and total recoverable petroleum hydrocarbons ranged up to 1350 and 5120 mg/kg, respectively.

Engineering soil tests, including particle size distribution analysis, density, dry unit weight, moisture content, and the Los Angeles Rattler test, were performed. The majority of the affected soil was coarse-grained, averaging 86.87% above sieve size No. 50. Particle size distribution analysis showed that PNAs were concentrated primarily in the relatively coarser-grained fractions (i.e., equal to or above No. 4 fraction), as illustrated in Figure 12-2. In addition, 47% of the soil matrix was demonstrated to be nonaffected and would not require processing or special handling. Sand equivalent averaged 36.33, exceeding the standard of 25 and thus considered nonexpansive. Density (weight) averaged 94.30 lb/ft^3, with an average moisture content of 12.13%. This equates to 2538.8 tons per cubic yard using a conversion factor of 1.26 ton/yd^3. The result of the Los Angeles Rattler test was Grade A (vs. B or C); thus, the soils were characterized as possessing good wear properties and resistance to abrasion.

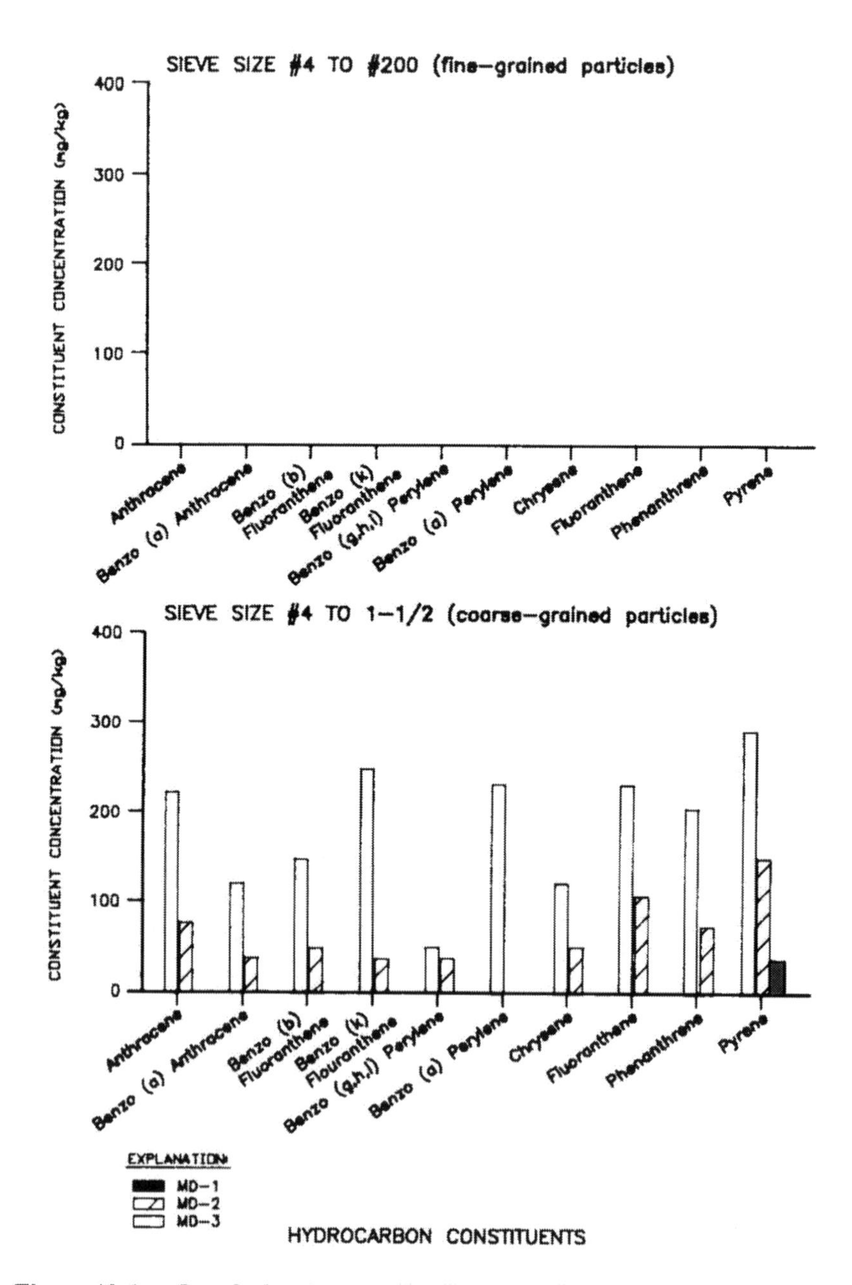

Figure 12-1. Graph showing results of particle size distribution analysis on PNA-contaminated soil generated at a former gas manufacturing plant.

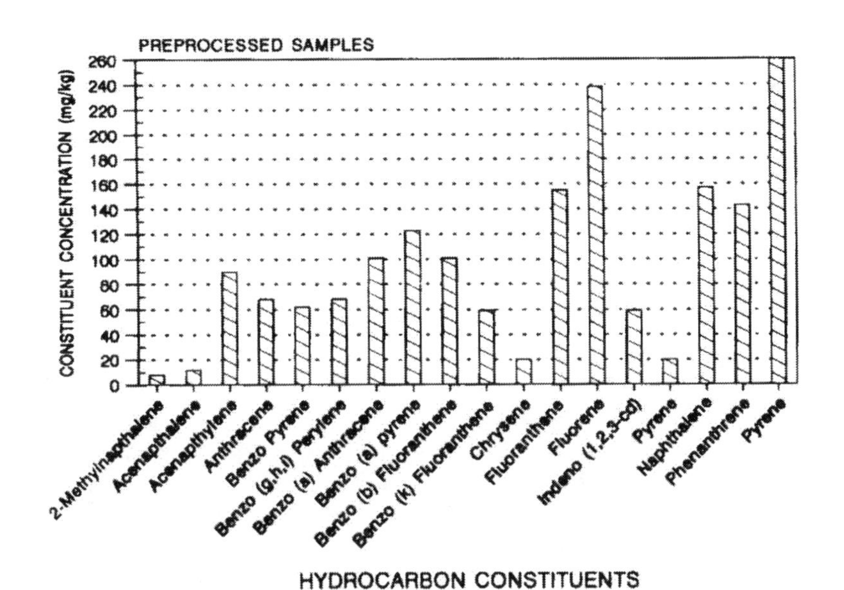

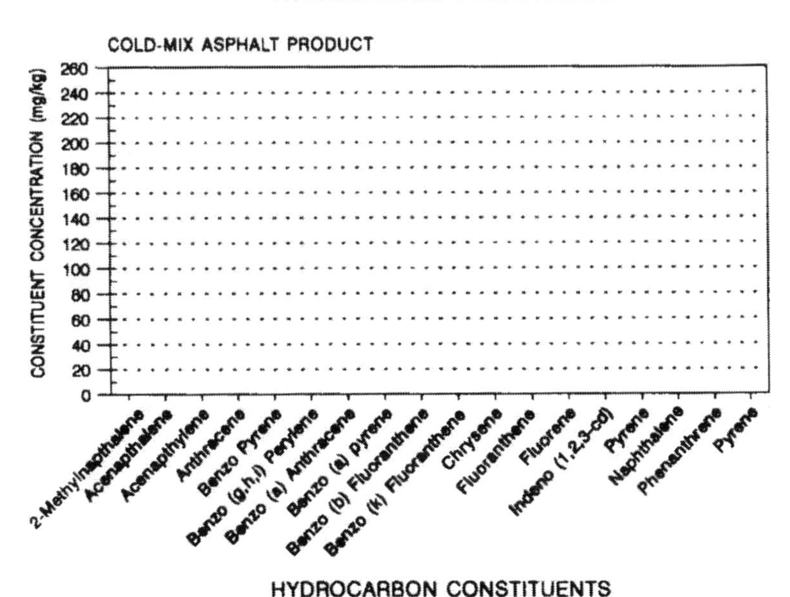

Figure 12-2. Graph showing analytical results on preprocessed PNA-contaminated soil and resulting processed CMA product samples generated at a former gas manufacturing plant.

Table 12-3 Summary of Engineering Test Results on Processed CMA Samples Incorporating PNA-Contaminated Soil

Mix design no.	Area (in.^{2})	Height after capped (in.)	Ultimate load (lb)	Load PSI	Unit weight (lb/ft^3)	Estimated Marshall stabilities[a] (psi)
MD-1	29.00	4.86	12,800	440	110.9	1100
MD-2	28.84	5.02	13,200	460	116.2	1150
MD-3	28.65	5.38	12,700	440	137.4	1100
MD-4	28.09	4.56	15,300	540	114.5	1350

[a] Approximate, based on a mathematical relationship between compressive strength and Marshall stabilities of 2.5:1.

Four mix designs were formulated (Table 12-3). Due to insufficient quantities of material to be tested, compressive strength testing was performed in lieu of Marshall stability. The relationship between compressive strength and Marshall stability strength is on the order of 2.5 to 1, respectively. Compressive strength (ultimate load) results, while providing an indication of stability, ranged from 12,700 to 15,300 lb. Thus, the objective of producing stabilities of 800 to 1200 psi was attained.

Chemical testing of the produced product for total petroleum hydrocarbons (EPA Method 8015 modified for PNA) PNA used EPA Method 25 and bio-aquatic assay LC50. EPA Method 625 was used in lieu of EPA Method 8270 since the medium being tested was leachate (water). All results were reported as nondetect, as is illustrated in Figure 12-1.

From a cost perspective, the cost of imported backfill could be reduced by about 47% (i.e., for a 30,000 yd^3 project with imported backfill delivered to the site, at a cost of about \$14/yd^3, the result would be a cost savings of \$197,400).

Chemical data for PNA-contaminated soil derived at another former gas manufacturing site also exhibited similar results. Eighteen various PNA compounds were identified in preprocessed soil ranging up to 91,000 mg/kg. Upon CMA processing, the resulting processed CMA samples were all reported as nondetect.

12.4 LEAD-MINE SCALE FROM FORMER GEOTHERMAL PLANT

Following demolition of a former geothermal plant, about 12,800 yd^3 of lead-mine scale-affected soil was encountered randomly disseminated throughout the upper 6 ft of silty clay soil. In addition to lead, elevated concentrations of arsenic, barium, and copper were found. Conduct of particle size distribution analysis on preprocessed soil showed the highest TTLC concentration for lead to be reported for the fine-grained fractions (i.e., equal to or less than No. 4 size fraction) as evident in Table 12-4 and as illustrated in Figure 12-3. The soil was characterized by a moisture content of 1.38%, with a sand equivalent of 8 (i.e., very expansive).

Chemical testing on these representative composited soil samples was performed (Table 12-5). Reported TTLC concentrations for lead in all of these samples exceeded 10× the STLC value of lead of 5.0 mg/l. One sample also had a reported TCLP concentration of 7.59, exceeding its respective TCLP MCL of 5.0.

Four mix designs were formulated (MD-1 through MD-4). The amount of lead-affected soil incorporated ranged from 45 to 75% by volume for mix design MD-1 and MD-4, respectively. Aggregate ranged from 17 to 47% in total volume, whereas emulsion remained essentially constant at 8 and 10%, respectively. In addition, a minor amount (less than 2%) of Portland cement was added.

Compacted product samples were tested via conventional Marshall stability procedures. Marshall stabilities ranged from 3260 to 4340 lb per unit weight of 115.0 and 116.0, respectively. After heating at 140°F for 2 h, Marshall stabilities values ranged from 810 to 1690 lb (Table 12-6). The unheated values significantly exceeded the minimum required valued at 2224 for paving mixtures.

Four CMA product samples were analyzed for TTLC for metals, and STLC and TCLP for lead (Figure 12-4). No elevated lead (or other metals) exceeding their respective MCL for TTLC was reported. STLC results were reported at 0.6, 0.7, 0.6, and 0.4 mg/l, none of which exceeded the STLC MCL for lead of 5.0 mg/l. In addition, all TCLP results were reported as nondetect.

12.5 LEAD- AND ZINC-CONTAMINATED SOIL FROM A BRAKE SHOES MANUFACTURING PLANT

The process of manufacturing cast iron brake shoes originated with imported raw iron ore and other ore materials, which were processed through an on-site foundry. The produced molten metals were poured into sand molds, quenched, and mechanically trimmed to approximate shape. Final dimensions were accomplished by machining and sand blasting. The sandblast grit resulting from this final dimensioning process was recycled on-site through a magnetic ore and water separation system. Large sumps, located within the complex, served as clarifiers that held the water and fines from the sandblast grit recycling system. The first step was to pump the clarifier water into vacuum trucks for disposal. The remaining material in the bottom of the clarifiers was characterized as fine-grained sandy residue, consisting of 90% sieve size #50, #100, and #200 minus particles, and 10% extraneous materials (i.e., wood, nuts, and bolts), with a moisture content of about 20%.

Preprocessed STLCs for lead, cadmium, and zinc were previously reported at 438, 1.42, and 336 mg/l.

Table 12-4 Summary of Analytical Results for Preprocessed Pb-Contaminated Soil and Processed CMA Product Samples

Parameter	Detection limit	Unit	Preprocessed soil			Processed CMA product				MCL[a]
			North	Central	South	MD-1	MD-2	MD-3	MD-4	
			Total concentration							(TTLC[a])
Antimony	3.0	mg/kg[b]	ND[c]	ND	ND	ND	ND	ND	ND	500
Arsenic	5.0	mg/kg	ND	ND	ND	ND	27	15	8.9	500
Barium	3.0	mg/kg	92.0	90.0	93.0	240	200	240	160	10,000
Beryllium	0.1	mg/kg	0.25	0.20	0.22	ND	ND	ND	ND	75
Cadmium	0.5	mg/kg	3.45	4.10	3.60	4.2	5.7	11	3.3	100
Chromium	1.0	mg/kg	8.20	10.1	7.90	13	38	14	11	500
Chromium (hexvalent as Cr+6)	1.0	mg/kg	ND	ND	ND					2500
Cobalt	2.0	mg/kg	3.60	4.40	4.10	10	12	12	8.6	8000
Copper	3.0	mg/kg	15.1	18.6	20.6	29	25	36	24	2500
Lead	5.0	mg/kg	145	221	137	210	350	230	160	1000
Mercury	0.05	mg/kg	ND	ND	ND	ND	ND	ND	ND	20
Molybdenum	3.0	mg/kg	ND	ND	ND	ND	ND	ND	ND	3500

Nickel	2.0	mg/kg	7.50	10.0	8.70	35	58	36	36	2000
Selenium	2.0	mg/kg	ND	ND	ND	ND	ND	ND	ND	100
Silver	1.0	mg/kg	ND	ND	ND	6	10	8.4	8.0	500
Thallium	5.0	mg/kg	ND	4.5	2.00	ND	ND	ND	ND	700
Vanadium	5.0	mg/kg	10.2	8.6	9.10	32	34	33	26	2400
Zinc	5.0	mg/kg	93.0	91.0	95.0	140	290	500	120	5000
		California waste extraction test (WET)								**(STLC[f])**
Lead	5.0	mg/l[e]	8.23	2.80	3.40	0.6	0.7	0.6	0.4	5.0
		Toxicity characteristics leaching procedure (TCLP)								**(MCL)**
Lead	0.2	mg/l	7.59	3.10	3.70	ND	ND	ND	ND	5.0

[a] MCL = Maximum Contaminant Level.

[b] mg/kg = milligrams per kilogram or equivalent to parts per million.

[c] ND = Not detected at concentration exceeding its respective detection limit.

[d] TTLC = Total Threshold Limit Concentration (CCR 66261.24(a)(2).

[e] mg/l = milligrams per liter or equivalent to parts per million.

[f] STLC = Soluble Threshold Limit Concentration (CCR 66261.24(a)(2).

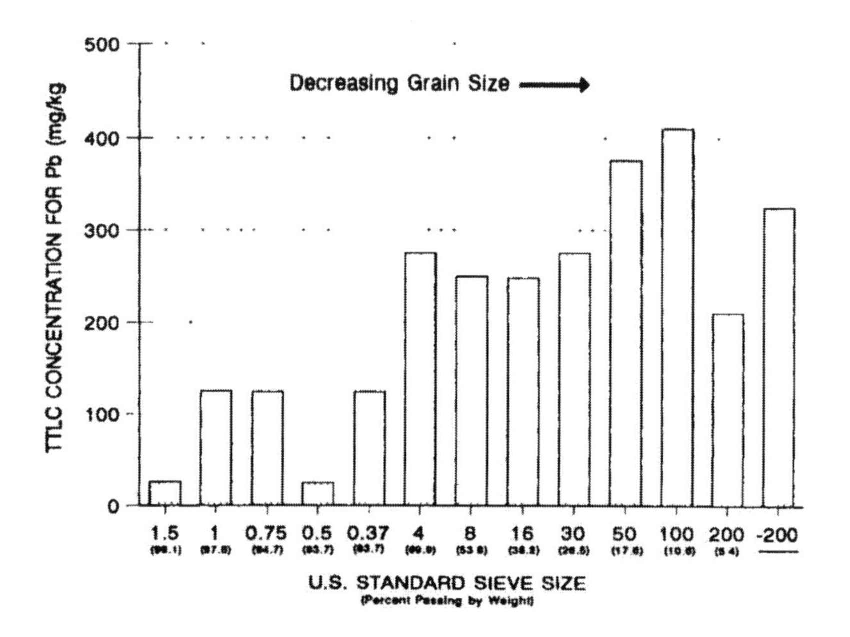

Figure 12-3. Graph showing results of particle size distribution of Pb-contaminated soil generated at a former geothermal plant.

Table 12-5 Particle Size Distribution for Lead

Size fraction	TTLC (mg/kg)	STLC (mg/l)	TCLP (mg/u)
1½	30.9	NA	NA
1	186	1.14	ND (<0.20)
¾	120	2.18	ND (<0.20)
½	27.3	NA	NA
⅜	119	0.74	ND (<0.20)
No. 4	276	1.49	NA
No. 8	246	1.88	NA
No. 16	237	2.84	NA
No. 30	277	4.00	NA
No. 50	359	3.09	NA
No. 100	413	3.81	NA
No. 200	235	3.15	NA
200 Minus	329	3.77	NA

Note: NA = not analyzed; ND = not detected at or exceeding the analytical detection limit as shown in parentheses; TTLC = total threshold limit concentration; STLC = soluble threshold limit concentration; TCLP = toxicity characteristics leaching procedure.

Table 12-6 Summary of Marshall Stabilities on Processed CMA Samples Incorporating Lead-Mine Scale-Contaminated Soil

Mix design no.	Marshall stability[a]		Unit weight (lb/ft^3)	Marshall stability after heating[b]	
	Average (lb)	Flow (in.)		Average (lb)	Flow (in.)
MD-1	3260	0.15	115.6	1090	0.14
MD-2	4050	0.17	116.6	1690	0.14
MD-3	4140	0.20	116.1	1050	0.17
MD-4	4340	0.30	116.0	810	0.29

[a] Cored at ambient temperature.

[b] Tested after heating at 140°F for 2 h.

Mix design was formulated at 44% metal-contaminated soil, 50% ¾-in. minus Class II aggregate, and 6% asphalt emulsion. Twenty-two tons of metal-affected soil were produced resulting in 50 tons of finished CMA product.

Product samples were analyzed using TCLP. Lead and zinc were reported at 3.97 and 7.62 mg/l, respectively, well below the TCLP for lead MCL of 5.0 and STLC for zinc of 250 mg/l (Figure 12-5). Cadmium was reported at nondetect. In comparison with preprocessed STLC values to processed sample values, lead showed a reduction ranging from 4.81:1 to 8.67:1. Zinc showed reductions ranging from 4.46:1 to 4.62:1.

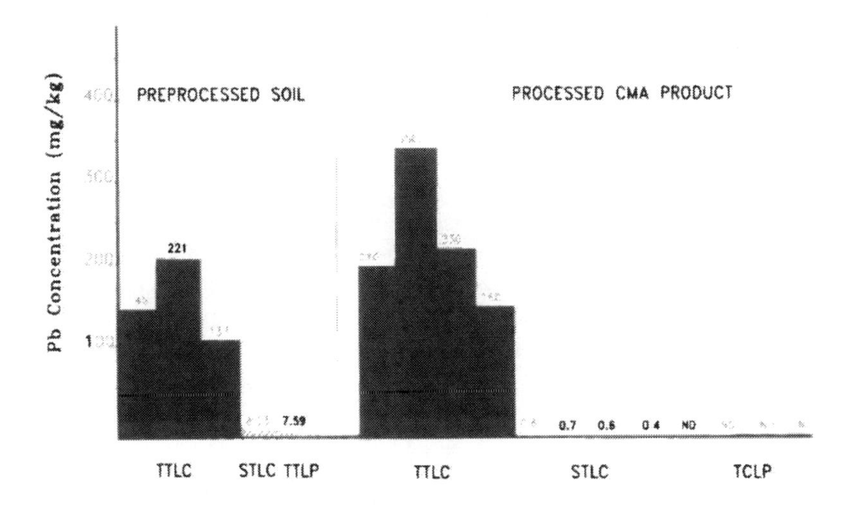

Figure 12-4. Graph showing analytical results of preprocessed Pb-mine scale contaminated soil and processed CMA product samples generated at a former geothermal plant.

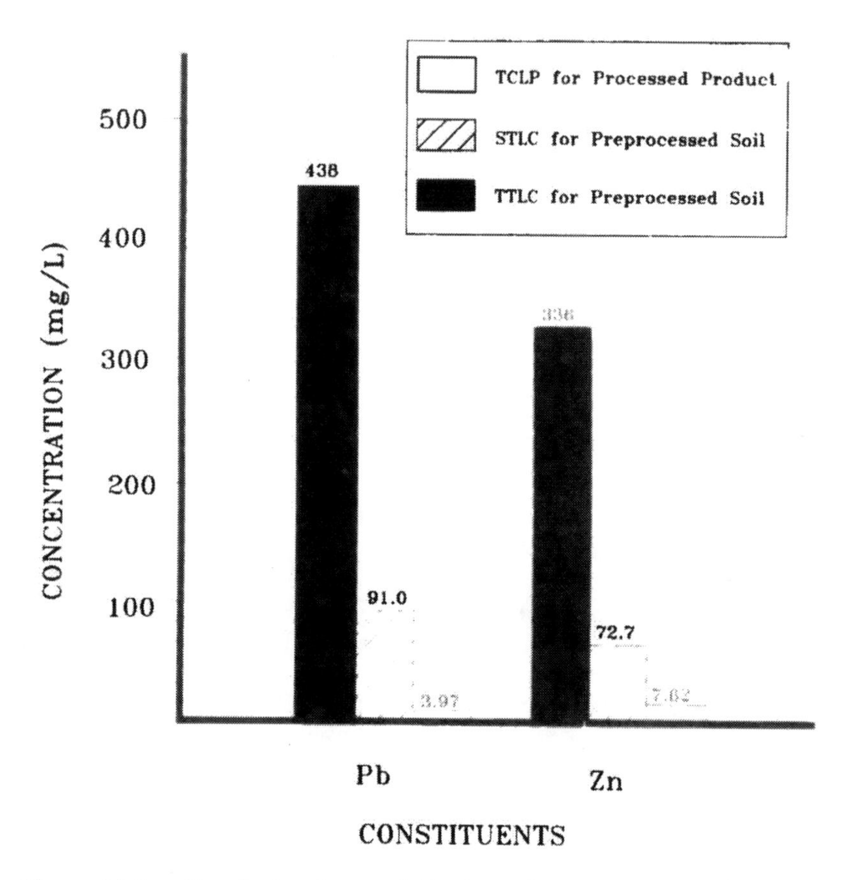

Figure 12-5. Graph showing analytical results of preprocessed Pb- and Zn-contaminated soil and resulting processed CMA product samples generated at a brake shoe manufacturing plant.

From a cost perspective, significant savings resulted (Table 12-7). Use of the end product saved at minimum $173/ton in state and county taxes. Comparative disposal cost savings were $110/ton. The finished product used by the client as pavement produced a $20/ton credit in comparison to the already low cost. Overall, this project saved the client a net $287/ton. Relative comparative costs for CMA incorporation, landfill, and incineration were $127, $266, and $1700 per ton, respectively.

12.6 LEAD- AND HYDROCARBON-CONTAMINATED SOIL FROM AUTO WRECKING SITE

During initial excavation of surface and near-surface soil, approximately 41 stockpiles of lead and hydrocarbon-contaminated soil was generated at an

Table 12-7 Cost Comparison for CMA Incorporation Vs. Hazardous Waste Landfill Disposal (HNLD)

Task no.	Description	CMA	Landfill disposal	Difference (per ton)
1.0	Sample and analyze affected material (waste characterization plus metals).	$3,200.00 L.S.	$3,200.00 L.S.	0
2.0	Excavate and load affected material.	$2,200.00 L.S.	$2,200.00 L.S.	0
3.0	Transport material. Tonnage rates for CMA material transport vary according to mileage and routing. (Price shown is site-specific.)	$3.25/ton	$37.50/ton	$34.25
4.0	CMA process vs. disposal of material. (Includes pretreatment per land disposal restriction regulations.)	$40.00/ton	$150.00/ton	$110.00
5.0	Value of finished product assigned by client. (Net after transportation and application as pavement at site of origin.)	$<20.00>/ton	0	$20.00
6.0	State taxes for disposal of hazardous waste.			
	Superfund landfill	0	$52.50/ton	$52.50
	Hazardous waste disposal fee	0	$105.00/ton	$105.00
	Generators fee and surcharge	0	$6.00/ton	$6.00
	County tax	0	$9.50/ton	$9.50
	Totals per ton (lump sum tasks no. 1 & 2)	$23.25	$360.50/ton	N/A
	Difference per ton CMA vs. HWLD	0	0	$337.25
	Additional analytical performed on the CMA finished product for the purpose of this project was $1100 or $50/ton.	$1,100.00	0	$50.00
			NET	$287.25

Note: Even more dramatic is CMA vs. incineration. The incineration facility fee alone is 78 to 90 cents/lb. At 90 cents/lb that is $1800/ton. For 22 tons, that equals $39,600. Then add $90 a ton for transport and taxes, and the incineration cost is well over $2000/ton.

Figure 12-6. Photograph of CMA processing of petroleum hydrocarbon- and Pb-affected soil generated at an auto wrecking site.

auto wrecking site (Figure 12-6). Excavated soil, derived from what was referred to as the auto smasher and dismantling areas, was transported to a former parking area and stockpiled on plastic. Each stockpile contained approximately the same volume of soil, totaling about 1394 yd^3. Two representative soil samples were retrieved from each stockpile for chemical testing prior to processing.

Once screened through a No. 2 in. sieve size to remove inert oversized materials, approximately 4009 tons of affected soil were generated. The screened soils were stockpiled based on whether they were characterized as hydrocarbon- or lead and hydrocarbon-affected. About 2673 tons of hydrocarbons-contaminated soil were segregated from about 1336 tons of lead-contaminated soil.

The soils were characterized as fine-grained silty sand with some gravel, with an average of 73% of the material passing the No. 4 sieve. Sand equivalent results indicated the soil to be nonexpansive, with a maximum density of 132.0 pcf at 8.4% moisture content. Actual in-place moisture contents were measured at 5.3, 4.0, and 7.3%, well below optimum moisture content. R-value results indicated that the soils met California Caltrans processed miscellaneous-base specifications.

Preprocessed soil had been reported with STLC and TCLP lead concentrations of 22 and 8.3 mg/l, respectively, exceeding the established TCLP MCL of 5.0 mg/l. Total recoverable petroleum hydrocarbons were reported to range up to 13,000 mg/kg.

Two mix designs were formulated. The amount of contaminated soil incorporated was 60 and 70% by weight, with added aggregate ranging from 32 to

22% by weight, respectively. The amount of water-based emulsion was kept relatively constant at 8%.

Representative hand-pounded cores of processed product samples were routinely tested during processing via Marshall stability at 1-, 3-, 7-, 9-, and 12-d intervals to determine optimum curing time. Average Marshall stabilities ranged up to 3750 psi, exceeding the standard established for freeway grade Type A HMA of 2400 to 2440 lbs. Moisture content remained consistent in all samples relative to the industry standard 3-d cure time optimum moisture for structural CMA. Marshall stability indexes of 1180, 2816, 3017, and 3750 lb, with an average Marshall flow of 0.29, 0.18, 0.24, and 0.20 in. for 3-, 7-, 9-, and 12-d cures, respectively. Average unit weights (density) ranged between 128.8 to 137.7 lb/ft^3.

Approximately 6539 tons of CMA were produced. Chemical testing was performed on ten samples, one for each day of processing, as stipulated by the lead agency. Total petroleum hydrocarbon was reported as nondetect in all samples analyzed. TCLP for lead was reported at concentrations ranging from nondetect to 1.1 mg/l, marginally above the analytical detection limit of 1.0 mg/l, and significantly below the regulatory MCL of 5.0 mg/l, as illustrated in Figure 12-7.

12.7 COST CONSIDERATIONS

In determining the overall cost benefits of reuse and recycling of contaminated soil into useful commercially viable end products, one can evaluate the

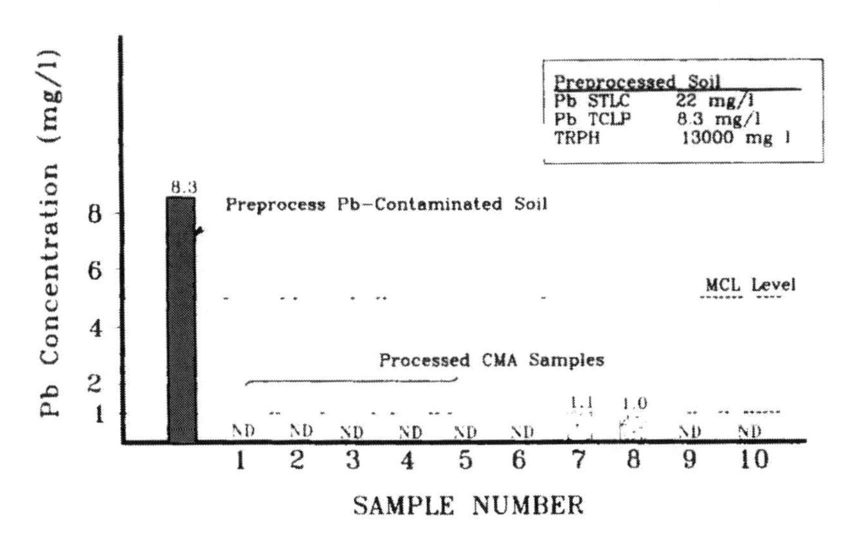

Figure 12-7. Graph showing analytical results of preprocessed Pb-contaminated soil and resulting processed CMA product samples generated at an auto wrecking site.

actual costs incurred and saved from projects formerly completed. For example, at a large municipal power generating station, about 2000 tons of oil contaminated soil are generated annually. About 700 tons of such soil were actually processed and used for pavement of a service road for dust abatement and heavy load-bearing traffic to a fuel oil storage area. Actual project costs were $91,000 at $130/ton (total turnkey including mobilization and demobilization, transportation, chemical and engineering testing and design, site preparation, and lay-down). Conventional land disposal costs would have been

Table 12-8 CMA Costs Vs. Disposal Costs for Non-RCRA Hazardous Waste in California

Task description	CMA costs[a]	Disposal costs[a]	Difference[a]
Site investigation	Same	Same	0
Characterization and workplan	Same	Same	0
Profiling and agency approval	Same	Same	0
Excavation	Same	Same	0
Transportation	0 for on-site processing	NA[b]	0
	$15.0/ton for central plant processing	$25.00/ton	$10.00
Disposal costs	NA	$65.00/ton	NA
vs.			
Processing costs	$35.00/ton (average)	NA	$30.00
Taxes and fees			
Superfund (HS)[c]	0	$7.88	$7.88
State taxes and fees	0	$26.25	$26.25
County tax 10% of disposal costs[c]	0	$6.50	$6.50
Site restoration[d]	Same	Same	0
Product cost recovery	$20.00/ton (average)	0	$20.00
		Total per ton	$100.63

[a] 1992 dollars.

[b] NA = Not applicable.

[c] State taxes and fees shown are from a recap of hazardous waste fees for fiscal year 1991–1992 compiled by the State Board of Equalization. Disposal costs shown above do not include generator fees or hazardous waste reporting surcharge.

[d] Site restoration costs where CMA was utilized on-site would be the same; however, the product cost recovery would be deducted from the gross site restoration costs, thereby providing a much lower actual net cost.

$140,000 at $200/ton. In consideration of the annual volume of 2000 tons, CMA incorporation (total turnkey) in comparison to landfill was $240,000 at $120/ton, compared to $100,000 at $200/ton, respectively, a cost savings of $160,000 at $120/ton. Actual processing costs, however, can be quite varied, ranging from as low as $25/ton to $45/ton. These costs are competitive with those associated with the most recent remedial technologies available, not even considering the added benefits of the value associated with the product, the tax benefits associated with the depreciation of the capital improvement, and the removal of liability as a generator of a hazardous waste. The cost-effectiveness of CMA is dramatic when compared to, for example, California's costs for disposal (Table 12-8). Where costs are noted as "same," this is to indicate that to utilize CMA technology, similar or lesser costs would be expended in comparison to the disposal option. The overall cost savings utilizing CMA technology vs. landfilling is on the order of $100/ton.

BIBLIOGRAPHY

Testa, S. M. and Patton, D. L., 1992, Add zinc and lead to pavement recipe, *Soils*, May, 1992, pp. 22–35.

Testa, S. M. and Patton, D. L., 1993, Resource recovery of lead-affected soil via asphaltic metals stabilization (AMS), in Proceedings of Superfund XIV, November 30–December 1, 1993, Hazardous Materials Research Control Institute, Washington, D.C., pp. 1219–1231.

Appendix A ACRONYMS

ANSI	American Nuclear Society Lead Test
APLT	Agitated Powder Leach Test Method
ARAR	Applicable or Relevant and Appropriate Requirement
ASTM	American Society of Testing and Materials
BUD	Beneficial Use Determination
CC	Coal Carbonization
CCR	California Code of Regulations
CERCLA	Comprehensive Environmental Response, Compensation, and Liability Act
CFR	Federal Code of Regulations
CMA	Cold-Mix Asphalt
CSA	Canadian Standards Association
CWG	Carburetted Water Gas
DAF	Dilution/Attenuation Factor
DLT	Dynamic Leach Test
ELT	Equilibrium Leach Test
EP	Extraction Procedure
HMA	Hot-Mix Asphalt
ILW	Intermediate Level Radioactive Waste
LLW	Low Level Radioactive Waste
kg	Kilogram
MAH	Monocyclic Aromatic Hydrocarbon
MEP	Multiple Extraction Procedure
mg/kg	Milligrams per kilogram or equivalent to parts per million
mg/l	Milligrams per liter or equivalent to parts to per million
MWEP	Monofilled Waste Extraction Procedure
NEC	Nuclear Energy Commission
NPL	National Priority List
OG	Oil and Natural Gas
PAH	Polycyclic Aromatic Hydrocarbon

PNA	Polynuclear Aromatic Hydrocarbon
PCB	Polychlorinated Biphenyl
RCRA	Resource Conservation and Recovery Act
SAP	Synthetic Acid Precipitation Leach Test
SARA	Superfund Amendments and Reauthorization Act of 1986
SCE	Sequential Chemical Extraction
SET	Shake Extraction Test
SET	Sequential Extraction Test
SLT	Soxhlet Leach Test Method
SLTM	Static Leach Test Method
STLC	Soluble Threshold Limit Concentration
TC	Toxicity Characteristics
TCLP	Toxicity Characteristic Leaching Procedure
TTLC	Total Threshold Limit Concentration
μg/l	Micrograms per liter or equivalent to parts per billion
USDOT	United States Department of Transportation
USEPA	United States Environmental Protection Agency
UST	Underground Storage Tank

Appendix B
GLOSSARY OF REGULATORY AND TECHNICAL TERMS

Provided below is a glossary of regulatory and technical terms pertinent to the reuse and recycling of contaminated soil. Terms are listed in alphabetical order, followed by a reference to each term's respective location as contained within the American Society for Testing and Materials (ASTM), American Association of State Highway and Transportation Officials (AASHTO); as the United States Code of Federal Regulations (CFR) by part or parts in parenthesis, followed by each term's definition.

Abandoned (40 CFR Part 261): Solid-waste materials that are (1) disposed of; (2) burned or incinerated; or (3) accumulated, stored, or treated (but not recycled) before or in lieu of being abandoned by being disposed of, burned, or incinerated.

Accumulated speculatively (40 CFR Part 261): A material that is accumulated before being recycled. A material is not accumulated speculatively, however, if the person accumulating it can show that the material is potentially recyclable and has a feasible means of being recycled, and that, during the calendar year (commencing on January 1), the amount of material that is recycled or transferred to a different site for recycling equals at least 75% by weight or volume of the amount of that material accumulated at the beginning of the period. In calculating the percentage of turnover, the 75% requirement is to be applied to each material of the same type (e.g., slags from a single smelting process) that is recycled in the same way (i.e., from which the same material is recovered or that is used in the same way). Materials accumulating in units that would be exempt from regulation under Section 261.4(c) are not to be included in making the calculation. (Materials that are already defined as solid wastes also are not to be included in making the calculation.) Materials are no longer in this category once they are removed from accumulation for recycling, however.

Aggregate (ASTM D 8-92): A granular material of mineral composition, such as sand, gravel, shell, slag, or crushed stone, used with a cementing medium to form mortars or concrete, or alone as in base courses, railroad ballasts, etc.

Anionic emulsion (ASTM D 8-92): A type of emulsion such that a particular emulsifying agent establishes a predominance of negative charges on the discontinuous phase.

Asphalt (ASTM D 8-92): A dark brown to black cementitious material in which the predominating constituents are bitumens, which occur in nature or are obtained in petroleum processing.

Asphalt cement (ASTM D 8-92): A fluxed or unfluxed asphalt specially prepared as to quality and consistency for direct use in the manufacture of bituminous pavements and having a penetration at 25°C (77°F) of between 5 and 300, under a load of 100 g applied for 5 s.

Asphalt rock (rock asphalt) (ASTM D 8-92): A naturally occurring rock formation, usually limestone or sandstone, impregnated throughout its mass with a minor amount of bitumen.

Asphalt rubber (ASTM D 8-92): A blend of asphalt cement, reclaimed tire rubber, and certain additives in which the rubber component is at least 15% by weight of the total blend and has reacted in the hot asphalt cement sufficiently to cause swelling of the rubber particles.

Asphaltenes (ASTM D 8-92): The high molecular weight hydrocarbon fraction precipitated from asphalt by a designated paraffinic naphtha solvent at a specified solvent-to-asphalt ratio.

Automatic proportioning control (ASTM D 8-92): A system in which proportions of the aggregate and bituminous fractions are controlled by means of gates or valves, which are opened and closed by means of self-acting mechanical or electronic machinery without any intermediate manual control.

Bank gravel (ASTM D 8-92): Gravel found in natural deposits, usually more or less intermixed with fine material, such as sand or clay, or combinations thereof; gravelly clay, gravelly sand, clayey gravel, and sandy gravel indicate the varying proportions of the materials in the mixture.

Batch plant (ASTM D 8-92): A manufacturing facility for producing bituminous paving mixtures that proportions the aggregate constituents into the mix by weighed batches and adds bituminous material by either weight or volume.

Bitumen-aggregate for recycling (ASTM D 4215-87) (Reapproved 1992): Bituminous pavement or paving mixture removed from its original location and reduced by suitable means, after removal or in place, to such particle size as may be required for use in cold-mixed, cold-laid recycled bituminous paving mixtures.

Bituminous (ASTM D 8-92): Containing or treated with bitumen (also bituminized). Examples include bituminous concrete, bituminized felts and fabrics, bituminous pavement.

Bituminous emulsion (ASTM D 8-92): (1) A suspension of minute globules of bituminous material in water or in an aqueous solution; (2) a suspension of minute globules of water or of an aqueous solution in a liquid bituminous material.

Blast-furnace slag (ASTM D 8-92): The nonmetallic product, consisting essentially of silicates and alumino-silicates of lime and of other bases, that is developed simultaneously with iron in a blast furnace.

By-product (40 CFR Part 261): A material that is not one of the primary products of a production process and is not solely or separately produced by the production process. Examples are process residues such as slags or distillation column bottoms. The term does not include a co-product that is produced for the general public's use and is ordinarily used in the form in which it is produced by the process.

Cationic emulsion (ASTM D 8-92): A type of emulsion such that a particular emulsifying agent establishes a predominance of positive charges on the discontinuous phase.

Clinker (ASTM D 8-92): Generally a fused or partly fused by-product of the combustion of coal, but also including lava and Portland cement clinker, and partly vitrified slag and brick.

Cold-mixed, cold-laid recycled bituminous paving mixtures (ASTM D 4215-87 (Reapproved 1992): Mixtures of bitumen aggregate for recycling with additional mineral aggregate as necessary, with or without mineral filler, mixed at or near ambient temperatures with additional bitumen.

Cold-mixed, cold-laid bituminous paving mixtures (ASTM D 4215-87) (Reapproved 1992): Mixtures of coarse and fine aggregates, or coarse or fine aggregate alone, with or without mineral filler, uniformly mixed and laid at or near ambient temperature.

Continuous mix plant (ASTM D 8-92): A manufacturing facility for producing bituminous paving mixtures that proportions the aggregate and bituminous constituents into the mix by a continuous volumetric proportioning system without definite batch intervals.

Crack filler (ASTM D 8-92): Bituminous material used to fill and seal cracks in existing pavements.

Cut-back asphalt (ASTM D 8-92): Petroleum residue (asphalt), which has been blended with petroleum distillates.

Cut-back products (ASTM D 8-92): Petroleum or tar residues, which have been blended with distillates.

Delivery tolerances (ASTM D 8-92): Permissible variations from the exact desired proportions of aggregate and bituminous material as delivered into the pugmill.

Dense-graded aggregate (ASTM D 8-92): An aggregate that has a particle size distribution such that when it is compacted the resulting voids between the aggregate particles, expressed as a percentage of the total space occupied by the material, are relatively small.

Discarded material (40 CFR Part 261): Any material that is (1) abandoned, as explained above, or (2) recycled, as explained below, or (3) considered inherently wastelike, as explained below.

Drum mix plant (ASTM D 8-92): A manufacturing facility for producing bituminous paving mixtures that continuously proportions aggregates, heats and dries them in a rotating drum, and simultaneously mixes them with a controlled amount of bituminous material. The same plant may produce cold-mixed bituminous paving mixtures without heating and drying the aggregates.

Dry mixing period (ASTM D 8-92): The interval of time between the beginning of the charge of dry aggregates into the pugmill and the beginning of the application of bituminous material.

Dust binder (ASTM D 8-92): A light application of bituminous material for the express purpose of laying and bonding loose dust.

Fine aggregate (ASTM D 2419-91): Aggregate passing the ⅜-in. (9.5-mm) sieve and almost entirely passing the No. 4 (4.75-mm) sieve, and predominantly retained on the No. 200 (75-μm) sieve.

Fog seal (ASTM D 8-92): A light application of bituminous material to an existing pavement as a seal to inhibit raveling, or to seal the surface, or both. Medium and slow-setting bituminous emulsions are usually used and may be diluted with water.

Fractured face (ASTM D 8-92): An angular, rough, or broken surface of an aggregate particle created by crushing, by other artificial means, or by nature.

Gas-house coal tar (ASTM D 8-92): Coal tar produced in gas-house retorts in the manufacture of illuminating gas from bituminous coal.

Generator (40 CFR Part 260.1): Any person, by site, whose act or process produces hazardous waste identified or listed in 40 CFR Part 261 or whose act first causes a hazardous waste to become subject to regulation.

Hazardous waste (40 CFR Part 260.1): A hazardous waste as defined in 40 CFR part 261.3.

Hot aggregate storage bins (ASTM D 8-92): Bins that store the heated and separated aggregates prior to their final proportioning into the mixer.

Incompatible waste (40 CFR Part 260.1): A hazardous waste, which is unsuitable for (1) placement in a particular device or facility because it may cause corrosion or decay of containment materials (e.g., container inner liners or tank walls), or (2) commingling with another waste or material under uncontrolled conditions because the commingling might produce heat or pressure, fire or explosion, violent reaction, toxic dusts, mists, fumes, or gases, or flammable fumes or gases.

Landfill (40 CFR Part 260.1): A disposal facility or part of a facility where hazardous waste is placed in or on land and which is not a pile, a land treatment facility, a surface impoundment, an underground injection well, a salt dome formation, a salt bed formation, an underground mine, or a cave.

Macadam, dry-bound and water-bound (ASTM D 8-92): A pavement layer containing essentially one-size coarse aggregate choked in place with an application of screenings or sand; water is applied to the choke material for water-bound macadam. Multiple layers must be used.

Maintenance mix (ASTM D 8-92): A mixture of bituminous material and mineral aggregate applied at ambient temperature for use in patching holes, depressions, and distress areas in existing pavements using appropriate hand or mechanical methods in placing and compacting the mix. These mixes may be designed for immediate use or for use out of a stockpile at a later time without further processing.

Maximum size (of aggregate) (ASTM D 8-92): *In specifications for, or descriptions of aggregate*, the smallest sieve opening through which the entire amount of aggregate is required to pass.

Mining overburden returned to the mine site (40 CFR Part 260.1): Any material overlying an economic mineral deposit that is removed to gain access to that deposit and is then used for reclamation of a surface mine.

Mixed-in-place (road mix) (ASTM D 8-92): A bituminous surface or base course produced by mixing mineral aggregate and cut-back asphalt, bituminous emulsion, or tar at the job-site by means of travel plants, motor graders, drags, or special road-mixing equipment. Open or dense-graded aggregates, sand, and sandy soil may be used.

Mulch treatment (ASTM D 8-92): A spray application of bituminous material used to temporarily stabilize a recently seeded area. The bituminous material can be applied to the soil or to straw or hay mulch as a tie-down, also.

Naphthene-aromatics (ASTM D 8-92): A mixture of naphthenic and aromatic hydrocarbons which are adsorbed from a paraffinic solvent on an adsorbent during percolation and than desorbed with an aromatic solvent such as toluene.

Native asphalt (ASTM D 8-92): Asphalt occurring as such in nature.

Nominal maximum size (of aggregate) (ASTM D 8-92): *In specifications for, or descriptions of aggregate*, the smallest sieve opening through which the entire amount of the aggregate is permitted to pass.

Oil-gas tars (ASTM D 8-92): Tars produced by cracking oil vapors at high temperatures in the manufacture of oil gas.

On-site (40 CFR Part 260.1): The same or geographically contiguous property, which may be divided by public or private right-of-way, provided the entrance and exit between the properties is at a cross-roads intersection, and access is by crossing as opposed to going along the right-of-way. Noncontiguous properties owned by the same person but connected by a right-of-way, which he controls and to which the public does not have access, is also considered on-site property.

Open-graded aggregate (ASTM D 8-92): An aggregate that has a particle size distribution such that when it is compacted, the voids between the aggregate particles, expressed as a percentage of the total space occupied by the material, remain relatively large.

Penetration (ASTM D 8-92): The consistency of a bituminous material expressed as the distance in tenths of a millimeter (0.1 mm) that a standard needle penetrates vertically a sample of the material under specified conditions of loading, time, and temperature.

Penetration macadam (ASTM D 8-92): A pavement layer containing essentially one-size coarse aggregate, penetrated in place by a heavy application of bituminous material, followed by an application of a smaller size coarse aggregate, and compacted. Multiple layers containing still smaller coarse aggregate may be used.

Pile (40 CFR Part 260.1): Any noncontainerized accumulation of solid, nonflowing hazardous waste that is used for treatment or storage.

Pitches (ASTM D 8-92): Black or dark-brown solid cementitious materials, which gradually liquefy when heated and which are obtained as residues in the partial evaporation or fractional distillation of tar.

Plant mix, cold-laid (ASTM D 8-92): A mixture of cut-back asphalt, bituminous emulsion, or tar and mineral aggregate, prepared in a central bituminous mixing plant and spread and compacted at the job-site when the mixture is at or near ambient temperature.

Plant mix, hot-laid bituminous emulsion mixtures (ASTM D 8-92): A mixture of emulsion and heated mineral aggregate, usually prepared in a conventional asphalt plant or drum mixer and spread and compacted at the job site at a temperature above ambient.

Plant screens (ASTM D 8-92): Screens located between the dryer and hot bins which separate the heated aggregates into the proper hot bin sizes.

Polar-aromatics (ASTM D 8-92): A polar aromatic hydrocarbon fraction that is adsorbed on an adsorbing medium from a paraffinic solvent during percolation and then desorbed with a chlorinated hydrocarbon solvent such as trichloroethylene.

Prime coat (ASTM D 8-92): An application of a low-viscosity bituminous material to an absorptive surface, designed to penetrate, bond, and stabilize this existing surface and to promote adhesion between it and the construction course that follows.

Pugmill (ASTM D 8-92): A device for mixing the separate hot aggregate and bituminous components into a homogenous bituminous concrete ready for discharge into a delivery vehicle.

Reclaimed (40 CFR Part 261): A material that is processed to recover a product if it is usable or if it is regenerated. Examples are recovery of lead values from spent batteries and regeneration of spent solvents.

Reclaimed asphalt pavement (RAP) (ASTM D 8-92): Asphalt pavement or paving mixture removed from its original location for use in recycled asphalt paving mixture.

Recycled (40 CFR Part 261): A material that is used, reused, or reclaimed.

Recycled asphalt paving mixture (ASTM D 8-92): A mixture of reclaimed asphalt pavement with the inclusion, if required, of asphalt cement, emulsified asphalt, cut-back asphalt, recycling agent, mineral aggregate, and mineral filler.

Recycling agent (RA) (ASTM D 8-92): A blend of hydrocarbons with or without minor amounts of other materials that is used to alter or improve the properties of the aged asphalt in a recycled asphalt paving mixture.

Reused (40 CFR Part 261): (1) A material that is employed as an ingredient (including use as an intermediate) in an industrial process to make a product (for example, distillation bottoms from one process used as a feedstock in another process). However, a material will not satisfy this condition if distinct components of the material are recovered as separate end products (as when metals are recovered from metal-containing secondary materials). (2) A material that is employed in a particular function or application as an effective substitute for a commercial product (for example, spent pickle liquor used as phosphorous precipitant and sludge conditioner in wastewater treatment).

Rubble (ASTM D 8-92): Rough stones of irregular shapes and sizes, broken from larger masses either naturally or artificially, as by geologic action, in quarrying, or in stone cutting or blasting.

Sand equivalent (ASTM D 2419-91): A measure of the amount of silt or clay contamination in the fine aggregate (or soil) as determined by test (see

Terminology D 653). (For further explanation, see Summary of Test Method and Significance and Use.)

Saturates (ASTM D 8-92): A mixture of paraffinic and naphthenic hydrocarbons that on percolation in a paraffinic solvent are not adsorbed on the adsorbing medium. Other compounds such as naphthenic and polar aromatics are adsorbed, thus permitting the separation of the saturate fraction.

Screenings (ASTM D 8-92): A residual product resulting from the artificial crushing of rock, boulders, cobble, gravel, blast-furnace slag, or hydraulic cement concrete, all of which passed the smallest screen used with the crushing operation and most of which passed the No. 8 sieve.

Sludge (40 CFR Part 260.1): Any solid, semisolid, or liquid waste generated from a municipal, commercial, or industrial wastewater treatment plant, water supply treatment plant, or air pollution control facility, exclusive of the treated effluent from a wastewater treatment plant.

Slurry seal (ASTM D 8-92): An application of a fluid mixture of bituminous emulsion, fine aggregate, mineral filler, and water to an existing pavement. Single or multiple applications may be used.

Soil (ASTM D 2419-91): Sediments or other unconsolidated accumulations of solid particles produced by the physical and chemical disintegration of rocks which may or may not contain organic matter (see Terminology D 653).

Soil aggregate (ASTM D 8-92): Natural or prepared mixtures consisting predominantly of stone, gravel, or sand which contain a significant amount of minus No. 200 (0.075-mm) silt–clay material.

Solid waste (40 CFR Part 261): Any discarded material that is not excluded by Section 261.4(a) or that is not excluded by variance granted under Sections 260.30 and 260.31.

Steel slag (ASTM D 8-92): The nonmetallic product consisting essentially of calcium silicates and ferrites combined with fused oxides of iron, aluminum, manganese, calcium, and magnesium, that is developed simultaneously with steel in basic oxygen, electric, or open hearth furnaces.

Stone chips (ASTM D 8-92): Small angular fragments of stone containing no dust.

Storage (40 CFR Part 260.1): The holding of hazardous waste for a temporary period, at the end of which the hazardous waste is treated, disposed of, or stored elsewhere.

Surface treatment (ASTM D 8-92): An application of bituminous material followed by a layer of mineral aggregate. Multiple applications of bituminous material and mineral aggregate may be used.

Tack coat (bond coat) (ASTM D 8-92): An application of bituminous material to an existing relatively nonabsorptive surface to provide a thorough bond between old and new surfacing.

Tar (ASTM D 8-92): Brown or black bituminous material, liquid or semi-solid in consistency, in which the predominating constituents are bitumens obtained as condensates in the destructive distillation of coal, petroleum, oil-shale, wood, or other organic materials, and which yields substantial quantities of pitch when distilled.

Tar concrete, cold-laid (ASTM D 8-92): A plant mix containing a medium-viscosity grade of tar and a graded mineral aggregate, designed to be laid either shortly after mixing or when the mixture is at or near ambient temperature.

Tar concrete, hot laid (ASTM D 8-92): A plant mix containing a high viscosity grade of tar and a densely graded mineral aggregate designed to be laid at or near the elevated temperature of mixing.

Transfer facility (40 CFR Part 260.1): Any transportation-related facility, including loading docks, parking areas, storage areas, and other similar areas, where shipments of hazardous waste are held during the normal course of transportation.

Treatability study (40 CFR Part 260.1): A study in which a hazardous waste is subjected to a treatment process to determine (1) whether the waste is amenable to the treatment process, (2) what pretreatment (if any) is required, (3) the optimal process conditions needed to achieve the desired treatment, (4) the efficiency of a treatment process for a specific waste or wastes, or (5) the characteristics and volumes of residuals from a particular treatment process. Also included in this definition for the purpose of the Section 261.4(e) and (f) exemptions are liner compatibility, corrosion, and other material compatibility studies and toxicological and health effects studies. A "treatability study" is not a means to commercially treat or dispose of hazardous waste.

Treatment (40 CFR Part 260.1): Any method, technique, or process, including neutralization, designed to change the physical, chemical, or biological character or composition of any hazardous waste so as to neutralize such waste or so as to recover energy or material resources from the waste, or so as to render such waste nonhazardous or less hazardous; safer to transport, store, or dispose of; or amenable for recovery, amenable for storage, or reduced in volume.

Treatment zone (40 CFR Part 260.1): A soil area of the unsaturated zone of a land treatment unit within which hazardous constituents are degraded, transformed, or immobilized.

Used in a manner constituting disposal (40 CFR Part 261): Materials are solid wastes when they are (1) applied to or placed on the land in a manner that constitutes disposal or (2) used to produce products that are applied to or placed on the land or are otherwise contained in products that are applied to or placed on the land (in which cases the product itself remains a solid waste).

Wet mixing period (ASTM D 8-92): The interval of time between the beginning of application of bituminous material and the opening of the mixer gate.

Appendix C
CONVERSION TO METRIC UNITS

To convert from	To	Multiply by
acre	meter^2 (m^2)	4 046.866
acre	hectometer^2 (hm^2)	0.404 686
barrel (42 gal)	decimeter^3 (dm^3) or liter (l)	158.987 3
BTU (International Table)	kilojoule (kJ)	1.055 056
Fahrenheit (temperature)	Celsius (°C)	$t_c = (t_f - 32)/1.8$
foot	meter (m)	0.304 80
foot^2	meter^2 (m^2)	0.092 903
foot^3	meter^3 (m^3)	0.028 317
	liter (l)	28.317 0
foot-pound-force	joule (J)	1.355 818
foot/minute	meter/second (m/s)	0.005 08
foot/second^2	meter/second^2 (m/s^2)	0.304 80
gallon (U.S. liquid)	decimeter^3 (dm^3) or liter (l)	3.785 412
	meter^3 (m^3)	0.003 785
gallon/minute	decimeter^3/second (dm^3/s) or liter/second (l/s)	0.063 09
gallon/yard^2	decimeter^3/meter^2 (dm^3/m^2) or liter/meter^2 (l/m^2)	4.527 314
inch	millimeter (mm)	25.400 0
inch^2	centimeter^2 (cm^2)	6.451 60
inch^2	millimeter^2 (mm^2)	645.160 0
inch^3	centimeter^3 (cm^3)	16.387 06
inch/second	meter/second (m/s)	0.025 40
inch of mercury (60°F)	pascal (Pa)	3 376.85
inch/second^2	meter/second^2 (m/s^2)	0.025 40
kilogram (kg)	ton (metric)	0.001 00
mile (U.S. statute)	kilometer (km)	1.609 344
mile^2	kilometer^2 (km^2)	2.589 988

To convert from	To	Multiply by
mile/hour	kilometer/hour (km/h)	1.609 344
minute (angle)	radian (rad)	0.000 290 89
ounce-force	newton (N)	0.278 013 9
ounce-mass	gram (g)	28.349 52
ounce-fluid	centimeter^3 (cm^3)	29.573 53
	liter (l)	0.029 574
pint (U.S. liquid)	liter (l)	0.473 176 5
poise (absolute viscosity)	pascal-second (Pa-s)	6.700 000
pound-mass	kilogram (kg)	0.453 592 4
pound-mass/foot^2	kilogram/meter^2 (kg/m^2)	4.882 428
pound-mass/foot^3	kilogram/meter^3 (kg/m^3)	16.018 46
	megagram/meter^3 (Mg/m^3)	0.016 018
pound-mass/inch^3	kilogram/decimeter^3 (kg/dm^3)	27.679.90
pound-mass/yard^2	kilogram/meter^2 (kg/m^2)	0.542 492
pound-mass/yard^3	kilogram/meter^3 (kg/m^3)	0.593 276
pound-mass/gallon	kilogram/meter^3 (kg/m^3)	119.826 4
(U.S. liquid)	kilogram/decimeter^3 (kg/dm^3)	0.119 826
psi	kilopascal (kPa)	6.894 757
quart (U.S. liquid)	decimeter^3 (dm^3) or liter (l)	0.946 352 9
ton (metric)	kilogram (kg)	1 000.000 0
ton (short-2000 lb)	kilogram (kg)	907.184 7
ton (long-2400 lb)	kilogram (kg)	1 016.046 1
ton-mass/yard^3	kilogram/meter^3 (kg/m^3)	1 186.552 7
yard	meter (m)	0.914 40
yard^2	meter^2 (m^2)	0.836 127 4
yard^3	meter^3 (m^3)	0.764 554 9

INDEX

A

B

C

F

G

H

N

O

Q

R

S

T

X

Y

Z